Synthesis Lectures on Image, Video, and Multimedia Processing

Series Editor

Alan Bovik, Cockrell School of Engineering, The University of Texas at Austin, Austin, TX, USA

This series publishes short books related to aspects of the imaging sciences that are relevant to furthering our understanding of the processes by which images, videos, and multimedia signals are formed, processed for various tasks, and perceived by human viewers. These categories are naturally quite broad, for two reasons. First, measuring, processing, and understanding perceptual signals involves broad categories of scientific inquiry, including optics, surface physics, visual psychophysics and neurophysiology, information theory, computer graphics, display and printing technology, artificial intelligence, neural networks, harmonic analysis, and so on. Secondly, the domain of application of these methods is limited only by the number of branches of science, engineering, and industry that utilize audio, visual, and other perceptual signals to convey information.

Stanislav Stanković

Game Design
for Free-to-Play Live Service

 Springer

Stanislav Stanković
Supercell Oy
Helsinki, Finland

ISSN 1559-8136   ISSN 1559-8144 (electronic)
Synthesis Lectures on Image, Video, and Multimedia Processing
ISBN 978-3-031-56155-9   ISBN 978-3-031-56156-6 (eBook)
https://doi.org/10.1007/978-3-031-56156-6

This Springer imprint is published by the registered company Springer Nature Switzerland AG
The registered company address is: Gewerbestrasse 11, 6330 Cham, Switzerland

Paper in this product is recyclable.

Preface

I fell in love with video games the very first moment I got a chance to play one for the first time. It was the 8-bit era of home consoles, and I was still too young to go to school. My uncle has burrowed a console from a friend of his for a day and brought it to our place to play with me. It was love at first sight. It was just like a cartoon, except I got to control the main character! A couple of years later, I got my first personal computer. This was the first time that I realized that I wanted to make games and not only play them. I had the rare fortune to have parents who were supportive of my gaming interests. They were both teachers of computer science at our local university and were wholeheartedly embracing the new age of technology. My mom taught me how to code, and my dad was bringing me games from his business trips. I will always be in debt to them for fostering my career.

Making games is a lifelong passion for me. However, my professional career is a little over a decade long. I had the audacity to send my CV and a link to a couple of games l made to an open application at Rovio's studio in the city of Tampere, Finland, where I lived at the time. I still don't know how, but by some magic, they decided to hire me as a game designer. This is where my journey began. It was a turbulent time when mobile games were a shiny new thing. Premium games still ruled the platform, but free-to-play was just around the corner. From Rovio, I continued my professional journey to EA and its Tracktwenty studio in Helsinki, Finland.

This book is a product of the experiences that I have accumulated during almost the 7 years I have been working at that studio. I had the good fortune to work on SimCity BuildIt, the latest incarnation of the venerable old SimCity franchise, one I had been playing as a kid in high school. This was also the time when the concept of free-to-play games evolved rapidly and reached its current mature form. It was also the period when the modern understanding of games as a service emerged. I had the rare privilege to work with a team of experts. Together, we have built something that we were truly proud of SimCity BuildIt is played by millions of people and has generated over half a billion dollars in revenue. At the time of writing these words, this game is still going strong.

At the moment, free-to-play games are the dominant model in the games industry. By some estimates, in 2021, these games generated over 100 billion dollars in revenue worldwide, which is far more than the movie and music industry combined. Even the franchises that were known for their premium titles are incorporating more and more service-like elements. The evolution of gaming, as well as the revolution of entertainment in general, is gravitating towards the service model.

This book is focused on the free-to-play games. In more broad terms, it is focused on games that are operated as a service. In other words, the games continue its life after being released to the public. Most of the examples I will be giving come from the world of mobile games, but almost all conclusions apply to all live games regardless of the platform. This book is dedicated to anyone interested in a deeper understanding of the game design problems of such games. This is not a game for the general public. I will delve quite deep into some topics, and a certain level of understanding of basic game design concepts and terminology is needed. Despite my best efforts, some parts of this book might seem tedious to the uninitiated.

Entertainment is a serious business. Game design is a complex field where various areas of human knowledge meet. Deep down, all games can be seen as mathematical systems. However, they are also much more than this. Game design is where mathematics and psychology come together to have a food fight. You can add to that a solid amount of business logic and visual and interactive art. This complexity of the topic is reflected in the organization of this book. Chapter 1 serves as an introduction in which we shall discuss the very basics, starting from the actual role of a game designer. In Chap. 2, we shall examine the business logic underlying free-to-play and game-as-a-service models. In Chap. 3, we are going to discuss video games from a psychological perspective, focusing on what makes certain activities feel fun and the underlying motivations of players. This knowledge about the psychology of individuals is applied in Chap. 4 to groups of people. In this chapter, we are going to discuss the idea of a target audience. Another facet of player psychology, the perception of time and the rhythm of games, is discussed in Chap. 5. The mathematics of games is discussed in more detail in Chap. 6. In Chap. 7, we examine the social glue that keeps gaming communities together. The process of designing a game is discussed in more detail in Chap. 8. Chapter 9 is dedicated fully to the problems of planning and running the live service. Finally, Chap. 10 consists of some practical game design examples.

This book is not meant to be a definitive book on this topic. Rather, it is a collection of experiences of one person. You could see it as a recipe book, something to start with, experiment, and build upon. I would be happy if this book served as a start of conversation if the material I present gives you some new idea or an impulse to move in a new direction. The field of game design is rapidly evolving, and so are the business models employed by the gaming industry. Many of the things presented in this book will become outdated in a matter of years if not months. I hope that the psychological and mathematical parts will remain. The human psyche and maths never change. Thank you for reading this book, and I hope that you will enjoy it.

Helsinki, Finland   Stanislav Stanković

Acknowledgments My son was born almost exactly a month before I started my first game designer job at Rovio. In him, I have a very visual reminder of how much my career in the industry grew. He is now a happy preteen whose life revolves around video games, much as mine was at that age. I would like to dedicate this book not only to him but also to my loving wife, my life partner with whom I have shared countless hours playing retro point-and-click adventures. I would also like to dedicate this book to my ever-supporting parents. Without their initial spark and support for my gaming interests, neither my career nor this book would ever be possible.

Contents

Introduction

What Is a Game?

Before we begin our exploration of game design, we need to ask ourselves one important question. What is a game?

This question can seem deceptively simple. After all, everyone, since the earliest age, intuitively knows what a game is. We instantly recognize a whole range of human activities as games. However, the definition in formal terms is much more difficult. It is precisely the broad scope of activities to which the term game is applied that makes defining it so challenging. What do a game of football, a game of chess, space invaders, Kimble, and poker have in common?

Over the last century, sociologists and anthropologists of various schools have offered a whole slew of definitions of this term, approaching it from multiple angles. See, for example, a definition by Ludwig Wittgenstein presented in his Philosophical Investigations (Wittgenstein, 1953), the one by the French sociologist Roger Caillois (Caillois, 1957), or even a more recent one by a game designer Chris Crawford (Crawford, 1984).

The definition that I prefer, however, is the one revolving around the notion of the *magic circle*. The term magic circle originates in Homo Ludens, a seminal book of Duch author Johan Huizinga (Huizinga, 1955), but it was popularised by Eric Zimmerman and Katie Salen in their more recent writings. See (Zimmerman et al., 2003).

In this framework, a magic circle is an invisible barrier separating the real physical world from the game world. A game world is a virtual space, a virtual environment in which the gameplay takes place. A game is any human activity taking place within the game world that is within the boundaries of the magic circle. Outside, in the real world, normal rules apply, the rules that govern everyday lives, including the laws of physics, manmade laws of human society, and even the unwritten cultural rules of decent behavior.

S. Stanković, *Game Design for Free-to-Play Live Service*, Synthesis Lectures on Image, Video, and Multimedia Processing, https://doi.org/10.1007/978-3-031-56156-6_1

Within the magic circle, a totally different set of *rules* applies. These rules would make little sense out of the context of a game. For example, twenty-two people chasing after a ball across a grassy playfield might seem absurd if it was not part of a football match. Others might even break the laws and conventions of normal life. Smacking another person's face with your fist is usually frowned upon unless it is part of a contact sport of boxing. Yet, within the context of the game, these rules make sense.

The idea of the set of rules as a defined aspect of a game is a common one, seen in many definitions of the term game. Indeed, as we soon see, it is fundamental for the work of a game designer.

Another important way of seeing a game is as a set of *goals* that the player tries to achieve and a set of *obstacles* that the player needs to overcome in order to attain these goals. The methods that a player can use to achieve these goals are defined by the rules of the games. The player's ability to overcome obstacles is governed by his skill and luck. We will devote a lot of space to the discussion of player goals and obstacles in this book.

Another topic that we are also going to discuss a lot in the following sections is the question of why people play games. Again, there are multiple ways in which this question can be answered. We are going to mention a couple of psychological theories dealing with human motivation, such as Self Determination Theory (Ryan & Deci, 2018) and the concept of the state of Flow (Csikszentmihalyi, 2008).

However, if we come back to the concept of the magic circle and the special set of rules contained within it, we realize the true power of games. These are tiny virtual worlds that act as psychological and sociological laboratories operating with reduced and deliberately constructed sets of rules that allow us humans to fulfill some of our basic psychological needs. In this way, they serve as a complement to our everyday lives, as a method of entertainment, and perhaps even healthy escapism (Hussain et al., 2021).

One of the fundamental human psychological needs is learning. Games are essential for learning (Hirsh-Pasek et al., 2004; Kahn & Wright, 1980). We begin to learn by playing games. We continue to learn through our lives even when we are not aware of it. This can include improving our physical skills, such as dexterity in handling a ball or a video game console, or learning about social interaction when playing with or against others. Arguably, many games cease to feel fun once one stops learning from them.

The following Chapters of this book are devoted to the discussion of all of these concepts in detail. However, keep in mind that this is just an approach to discussing and designing games. There are plenty of other very good books on this subject, for example, *Game Design Workshop: A Playcentric Approach to Creating Innovative Games* by Tracy Fullerton (Fullerton, 2018), *A Theory of Fun for Game Design* by Raph Koster (Koster, 2013), or *The Art of Game Design: A Book of Lenses*, by Jesse Schell (Schell, 2019).

The Role of the Game Designer

I grew up with the notion that video games are made. Video games are, after all, pieces of software. It only makes sense that the guys capable of writing the code are the ones calling the shots. Of course, I had a vague idea that there must be other professions involved, such as artists making the visuals, but it was only years later that I discovered that there is such a thing as a game designer.

Now, after a decade in the game industry working as a game designer, I still often find myself in the position to explain to people what it is that I actually do. Typically, if I begin by saying that I make games, I need to point out that I do not actually write any code. If I mention my job title, the word designer leads people to believe that I do some sort of visual design, i.e., produce graphics for the game.

The profession of a game designer is something that has evolved from the needs of the discipline. The first games were indeed originally created by programmers and coding enthusiasts. In many contemporary indie projects, the lead designer is also the lead programmer and a lead artist and multiple other roles. Therefore, the misconceptions that I described are not totally unjustified.

In reality, the work of a game designer is focused on that intangible in-between which binds both the code and the artwork into a coherent whole. A comparison of a game designer to a movie director or a scriptwriter is often a helpful one, as it brings to attention the unseen underlying structure of the game. A script is a blueprint for a movie. The work of multiple people is what brings it to life. The movie director is the one who coordinates all of these efforts to produce a coherent result.

However, this comparison can also lead people on the wrong path. Only some types of games have linear structures or try to tell a single narrative. A game design document resembles a blueprint, but it is not exactly the same thing as a scenario for a film. Instead of a narrative, games operate as systems of rules. One of the simplistic ways of explaining what we do is to say that we make up the rules of the game. Everyone is familiar with the rules of basketball, but someone had to specify that each team has exactly five players. Someone had to specify how long each quarter lasts and what is the total time allocated for a match. Someone had to specify how many points a team gets each time a player manages to score, etc.

Even within a game development team, the role of a game designer can sometimes be misunderstood. A part of the job is coming up with ideas, sometimes random new out-of-the-blue ideas for new features and gameplay patterns. More often, a game designer is expected to come up with solutions to specific problems.

Liz England, in her famous essay (England, 2014), uses the metaphor of a door problem to explain what the job of a game designer entails. She begins by asking a simple question, "Are there doors in your game?" and continues by expanding on it with a series of other questions: "Can the player open them?" "Can the player open every door in the game?". Each following question probes the new layer of detail, building upon the answers provided by the previous questions: "Does a player know how to unlock a door? Do they need a key? To hack a console? To solve a puzzle? To wait until a story moment passes?" etc.

I find this essay to be a very fitting explanation of a game design process. What we actually do is provide answers to design problems. Each new answer adds a new level of detail and usually opens up a new set of questions. Therefore, the design process is often open-ended. A shipped game, on the other hand, needs to be, at last, somewhat complete. Tying up loose ends is actually one of the key responsibilities of a game designer.

Quite often, game design is not about coming up with ideas. Ideas can come from any source, other team members, or even the players themselves. The job of a game designer often means collecting ideas from all those various sources, filtering them, refining them, and making them fit the rest of that big jumbled mess called a game. This becomes even more evident in the games that run as services and evolve over the years. Henri Roth, my first boss and mentor in this craft, used to compare the job of a game designer to a janitor. Our work becomes visible only if we do our job badly and we leave some mess behind.

The remainder of this book will focus on many aspects of this process. It can, indeed, entail coming up with the rules of the games and the underlying structure of the game. Constructing this structure can also be seen in creating goals that the player needs to reach and obstacles that the player needs to overcome in order to win or finish the game. In a way, the game design can be seen as structuring the player's time or playing with the player's motivations.

In the same way that the role of a dedicated game designer has evolved as games have evolved in complexity, various sub-disciplines have emerged to focus on various aspects of game design (England, 2014). As the name implies, *Level Designers* specialize in the craft of making individual levels in various level-based games. *Narrative Designers* focus on what most closely resembles scriptwriting in the world of gaming, on things like the lore, backstory, world-building, and character design. *System Designers*, on the other hand, tend to focus on seeing games as a set of interconnected rules typical for intricate and very deep games such as various simulations, city builders, strategies, or role-playing games. *Economy Designers* and *Monetization Designers* are specializations that have emerged with the arrival of free-to-play games and focus on the particular needs of this business model.

As video games continue to evolve, new profiles of game designers are sure to emerge.

Psychology and Mathematics

I am an engineer by training. Most people familiar with my educational background, on hearing about my profession, will immediately ask me about *Game Theory*, which is a specific area of mathematics designed specifically as a tool for the analysis of various games (Morgenstern & von Neumann, 1944).

Indeed, as sets of rules, games can be seen as mathematical systems; this is obvious in the case of well-known and much-studied games such as Chess. The importance of mathematics is also obvious in the case of games of chance and various gambling games, such as roulette, as well as various other card games. Still, even the most complex virtual worlds, such as World of Warcraft or EVE Online, can be described in mathematical terms. Mathematics is the description language of the universe. Chapter 6 of this book is devoted entirely to some practical examples of mathematical models, which I found useful in my game design practice.

However, mathematics provides one side of the story. Game theory tries to describe the most rational, the most beneficial set of actions that a player can make based on the available information within the context of the game. Video games are designed to be played by humans. Human players are anything but rational. We do strive to make the best possible choices at any given moment in time. However, our minds are not mathematical problem-solving machines. Our decision-making is warped by things like cognitive biases, instincts, and emotions.

The way in which humans perceive various mathematical systems can be surprising. Two systems that behave very similarly in the mathematical sense or even look almost the same can be perceived in quite different ways on the emotional level, as illustrated by the example in Fig. 1.1, first mentioned in a tweet by Cheeseeister (Cheesemeister, 2022).

Fig. 1.1 Super Mario Bros World 8 level 1 and level 3

Please do not leave with the impression that human brains are somehow flawed. Our cognitive biases have deep evolutionary roots (Haselton & Nettle, 2015). They exist because they have a positive impact on the chances of survival in the ever-changing universe.

Games are where mathematics and psychology meet. This is one thing that makes them so fascinating and compelling. Understanding the way in which the human psyche operates and how the human brain perceives mathematical systems is the key to designing entertaining and engaging games. We are going to discuss player psychology in detail in Chaps. 3 and 4.

Core Gameplay and Metagame

In the following chapters of this book, we are going to touch on many game design concepts. A lot of them we are going to discuss in great detail. Some of them, such as *game rules*, *player's goals*, and *obstacles* we have already mentioned. However, there are two additional concepts I feel that I still need to mention because they are essential for understanding the discussions that follow. These are the notions of the *core gameplay* and the *metagame*.

So far, we have talked about games as if each individual game is a monolithic piece of design, one unified whole, a coherent set of gameplay rules. This might be true for some very simplistic games. Most games, however, consist of various layers of gameplay that somehow interact.

To understand what I mean, let's take a relatively simple example of a very well-known game series, Super Mario Bros. All of these games are archetypal platformers, meaning that the player's objective in the game is to reach the end of the level by making the main character move rightwards and jump over various platforms and chasms in the scenery. This is the *core gameplay* of Super Mario Bros.

However, the game is not one single extremely long level, rather it is organized in a series of levels accessible via a separate screen, conventionally known as overworld. Furthermore, multiple levels are grouped into distinct larger units known as worlds. Each of these worlds is represented by its own overworld map. In addition, there is a whole set of rules related to lives, various powerups, etc. Finally, there is an overarching narrative of saving the princess that provides the backstory and the setting for the whole game. Everything beyond the core gameplay within one level of Super Mario Bross game is its *metagame*.

The metagame and the core game are closely connected. The core gameplay is embedded in the metagame structure. In many video games, the core gameplay is also known as *minute-to-minute* interaction. In some games, however, it is very hard to tell where the core gameplay ends and the metagame begins. This is typically the case in many city simulation games, such as SimCity, or turn-based strategy games, such as Civilization.

The distinction between the core and the metagame does not exist exclusively in the world of video games. In the game of chess, the core gameplay is, obviously, everything that takes place during a single match on an 8×8 chessboard. On top of this, there exists a whole set of tournaments, competitions, FIDE rankings, etc., that collectively constitute the metagame of chess.

Notice also that the metagame can transcend the structures that were explicitly created by the designers of the game. For a video game, parts of the metagame can also be all sorts of discussions on forums and various websites, YouTube and Twitch streams, Reddit subreddits and local e-sports competitions, fan art and mods, etc. Big chunks of the metagame can be out of the control of the original game designers, but in a way, a rich and prosperous fan-created metagame is a sign of success.

When it comes to Free-to-Play games, the type of games that this book is primarily aimed at, metagame is of extreme importance. Typically, the metagame is part of the game where the features and structures aimed at engaging and retaining players for a long time reside. This is usually the part of the game that ensures the longevity and profitability of the game as a service. Therefore, most of the discussions in the following chapters of this book are going to revolve around the metagame in one way or another.

Business Model

2

Market Landscape

The human brain is a remarkable thing. It is capable of doing things ranging from mundane such as coordinating bipedal locomotion to pondering the secrets of subatomic particles. Furthermore, it is capable of doing both things in parallel. However, one thing that it seems not to be able to handle well is free time. As humans, we have evolved in an everchanging environment beset with dangers and opportunities on all sides. Evolution has shaped our minds in such a way that we feel imperative to make productive use of every waking moment. The feeling of malaise that we sense when our mind is forced to sit idle is what we call boredom... and it can be unbearable!

Over the eons, humans have invented a myriad of ways to occupy our minds with something, anything in these moments of boredom, from storytelling around the campfire to instant messengers and immersive VR experiences. We call these things entertainment. It is the third decade of the 21st century and entertainment has many forms, film, TV, music, sports, and of course, interactive electronic entertainment, a.k.a. video games.

Entertainment is also a huge industry. For example, the global box office revenue in 2022, the year in which the movie industry was recovering from the disastrous COVID-19 times, was over $25.9 billion (Mitchell, 2023). The numbers posted by the global music industry for the same year are even greater, with over $31.5 billion in revenue (Mulligan, 2023).

Very few people are aware that these numbers are simply dwarfed by the video games industry. According to expert estimates, the cumulative global revenue of the global revenue by video game companies in 2022 was between $184.4 billion (Wijman, 2022) and $396.2 billion (Clement, 2023).

About half of this revenue or $92.2 billion came from mobile games, with console and PC games contributing an additional $51.8 billion and $38.2 billion respectively (Wijman,

© The Author(s), under exclusive license to Springer Nature Switzerland AG 2024
S. Stanković, *Game Design for Free-to-Play Live Service*, Synthesis Lectures on Image, Video, and Multimedia Processing, https://doi.org/10.1007/978-3-031-56156-6_2

2022). These revenue numbers are followed by a healthy compound annual growth rate (CAGR) estimated at 4.5%. About 95% of revenue on mobile and about 32% of revenue on console and PC platforms was generated by Free-to-Play games, making this by far the most dominant business model in the industry.

Video games are truly a global business. Asia and the Pacific contribute to the lion's share of the revenue with 48% followed by North America with 26% and Europe with 18%.

These huge revenue numbers originate from the popularity of video games as entertainment. According to the same estimates, there are 3.2 billion players worldwide. Not surprisingly most of them are located in Asia and the Pacific, with over 1.7 billion or 54%, followed by 511 million players in the Middle East, 428 million in Europe, 316 million in South America, and 232 million in North America.

Premium Games

What is now known as premium games was historically the second business model employed by the games industry. You might be surprised by this statement, but remember that the video game industry started with in 1972, with Atari and Pong its first arcade video game (Kent, 2001). The original business model of arcade video games was much more akin to the current Games-as-a-Service (GaaS) model (Bycer, 2018) than one would expect. It was not until the early 1980s and the arrival of the 8-bit era of gaming consoles that the premium model become the dominant way of distributing video games (Wolf, 2012).

The state of technology at that age was the key factor behind the rise of this trend. Personal computers and home gaming consoles allowed people to enjoy games from the comfort of their homes. Thus, much more often than they would if they needed to visit a dedicated establishment, i.e. a nearby arcade. On the other hand, at that time pretty much the only way to distribute software was to record it on some physical medium. Early consoles relied on dedicated cartridges (Edwards, 2015), and PCs had their floppy disks (Pace, 2022). Somewhere in the '90s industry moved to optical media, CDs and DVDs, etc.

These two factors shaped the way games started to be produced. Furthermore, they influenced the way in which game designers started to think about games. Video games began to be seen as products that could be packaged, just like any other type of physical goods, stamped onto the physical media and shipped to stores where they would be sold to customers.

The version of the game that would be recorded not the medium represented a finished product, a finished work of design. Designing a game became a form of art similar to writing a novel or making a feature film. The team working on a game had, more or less, total control over the player's journey, from the beginning until the end.

Physical goods, distributed in physical, brick-and-mortar stores are bound by the same constraints as other physical goods. They are distributed via the same, relatively slow logistic chains. In stores, each new title had to compete for the limited shelf space with all the other games. This in turn created marketing strategies that would focus on the moment when a new game is launched, creating a marketing hype that would produce an initial sales push. Indeed, most titles from this era derived most of their revenue based on the sales during the first two weeks after the initial launch.

To avoid the limitations of physical retail space, gaming companies gradually switched to digital distribution channels, i.e. method by which a game is downloaded directly onto computers. The origins of this idea surprisingly far back in time. Back in the 80s, people could get gems by recording radio station broadcasts onto audio cassettes (English, 2020)! Later on, in the 90s, the so-called shareware model of distributing games over BBS and the early Internet become popular (Edwards, 2021). However, digital distribution became a viable business model only with the arrival of broadband internet in the last couple of decades (Mattioli, 2020), with the advent of platforms such as Steam (Yu, 2017). Premium monetization model even dominated the first wave of mobile gaming on smartphones, all the way from the launch of the iOS AppStore in 2008 until about 2014, when Free-to-Play started to dominate this segment of the market.

It should be noted that the distribution of games on a physical medium has one very important advantage for the consumer and a significant disadvantage for the gaming companies. The consumer remains the owner of the physical game, free to resell it to other people. This in turn dramatically reduces the total cost of games purchased by the consumer, but also reduces the total revenue for the gaming industry, as only a fraction of people who play the game are paying the full price of the game. Furthermore, the difference in price often got pocketed by the middlemen facilitating this secondhand market, i.e. retail companies such as GameStop (Alexander & Matthews, 2009). This was an additional big reason behind the push for digital distribution by major game companies.

As we have mentioned in the previous section, the premium model of monetization is still quite relevant even today, contributing a big chunk of overall game industry revenue (Wijman, 2022).

The very nature of the premium model has some very important implications on the game design. Above all and by definition the financial transaction takes place before the consumer, the player has a chance to actually experience the game! This can be seen as a sort of investment. The consumer is making an investment of a certain amount of money in the hope that he will derive a certain amount of entertainment from the game.

How many times did you buy a game that disappointed you? That you abandoned mere hours or worst minutes into gameplay? From the financial perspective, this is when you didn't get your money's worth of fun. In order to minimize this risk, consumers are forced to rely on external sources of information about the game. In the olden days, this used to be word of mouth, but it quickly evolved into an industry of its own (Perreault & Vos, 2019). This is why game reviews and game journalism is so important for the premium business. Early game reviews can make or break sales of a new title. It is also the underlying reason for the continuous hostile attitude that the vast majority of game journalists have toward other monetization models, primarily Free-to-Play.

The visual impressions matter in this framework. Game visuals are usually the centerpiece of the marketing campaigns for premium games. This was also one of the hidden factors driving the further development of video game graphics, motivating in turn technological research (Sawicki & Moody, 2020).

Software Piracy

As you can tell from my biography, I was born and raised in Serbia. I spent my teenage years pretty much playing games. The world of gaming was in its 16-bit era. At that time Serbia was under heavy UN sanctions and in any case we were a poor country. Imported licensed games were mostly out of our budget, but copying games to flippies was ridiculously easy. So-called pirate shops proliferated around the country. Those guys would get games from god knows where. They would bring them to our town and they would make copies, onto your floppies, for a fixed price per floppy.

Eventually, CDs come along, but the first CD-ROM drives were read-only, and buying a special device able to record was ridiculously expensive. Luckily some smart soul figured out a way around this problem. For a while so-called CD-rip games become popular. CD-rip meant, that the game was stripped of bits of content, to make it fit on as little memory space as possible. In practice this meant, stripping off any audio and video files, basically all of its multimedia content, but who needs cutscenes anyway? You just want to play the game, right?

Even stripped like that games were still growing in size, far transcending the capacity of a single floppy disk. I remember going to our favorite pirate shop carrying two plastic shopping bags full of floppies. The pirate guy would already have the game zipped and chopped in files big enough to fit on a floppy. Recording that still took some minutes per floppy, so most likely you would spend quite some time at that place watching the guy swapping floppies. Finally, when you would get home, you'd have to do the reverse process and unzip all those files back to your PC's hard drive. Floppies were also kind of expensive, so naturally we were reusing most of them hundreds of times. This meant that some of them would eventually fail. Of course, you would discover this only when you get home. Oh, joy. This meant that you would have to go back to the pirate shop, with a different bunch of floppies and a list of zip files that you were missing. The pirate guy would then record files number 3, 6, 8, 16, 18, and 25, again onto your floppies. Luckily, the deal was that they wouldn't charge you again for those broken files. You'd return home in the hope that this time around nothing got broken, but there were no guarantees of course. I am pretty sure that for some stuff I had to make three trips.

Finally one of the pirate shops bought a CD recorder. It was actually a mutual investment of two prominent local pirates. The machine was finicky and the process was delicate and slow. The blank CDs were again kind of expensive. Any shaking or surprise vibrations would cause the process to fail and permanently damage the blank CD. Oh, the horror. To prevent this, the two entrepreneurs decided to place the device, an external thing in a sizable beige box, onto a separate desk. It was connected with a cable to a PC on the other desk. They were really proud of their investment so they parked the desk in the middle of their shop. The desk was a bit shaky and their young customers had a habit of leaning on it, which would in turn occasionally cause the burning process to bork the CD.

To solve the problem they placed a big handwritten sign next to the device saying DO NOT LEAN AGAINST THE DESK! The sign was duly ignored by most of the visitors, so the typical scene at that place was that a new customer would enter the shop, and casually take a step towards the desk. At that moment all of the employees in the shop would shout in unison: "DO NOT LEAN AGAINST THE DESK!".

Software piracy is not a topic that is often mentioned in game design books. Of course, it is not a business model by itself, however, it definitely is a sort of distribution model for many games. Historically, it had serious implications for the gaming industry. It is hard to estimate the amount of actual revenue loss caused by piracy. Only a fraction of people buying pirated games would be able to afford even some of those games at the full price (Ende et al., 2014). On the other hand, piracy is still existing and still hurting the sales of premium games (Goff, 2022). You can still find pirated versions of pretty much any premium software or game. Millions of people still use this stuff (Quintais & Poort, 2018).

Problems of software piracy were one of contributing factors driving the move towards the GaaS (Game as a Service) and the Free-to-Play business models. Having your game rely on a server-based backend and having the server constantly validate players' actions prevents a lot of problems related to software piracy. Distributing your game for free undercuts the profits of any pirate. This in turn has caused a dramatic shift in piracy trends. The piracy incidents declined by over 50% in the period between 2017 and 2020, and the majority of the pirated content, about 80%, is now related to streamed video content (Garcia-Valera et al., 2021).

Software piracy was always driven by two factors, price and availability. Piracy in the 90s and early 2000s made premium games and software, accessible to a much bigger audience of people around the world. Back in the day, gaming companies tended to focus most of their attention on a couple of key markets of the global North and West. Usually to places with developed distribution networks and populace with enough disposable income to be able to afford their products. This left billions of people out of the loop. Once games started to become widely available via digital distribution and started to be available for free.

I can safely admit that I owe my career to piracy. I definitely owe my encyclopediac knowledge of 16-bit era PC games to this.

Free-to-Play Games

In this section, we finally turn our attention to the business model that this book will focus on the most, i.e. Free-to-play (F2P) games and its closely related concept of Game-as-a-Service (GaaS).

Compared to premium games, Free-to-Play is a much newer business model. It emerged in the late 90s out of the community around Massive Multiplayer Online (MMO) games. The exact origins are hard to pinpoint, and some sources credit Matt Mihaly, the CEO of Iron Realms Entertainment as implementing the first instance of it inAchaea, Dreams of Divine Lands their text-based Multi-User Dungeon (MUD) in 1997 (Hrodey, 2020).

The concept became truly popular in East Asia, especially in Korea at the start of the 2000s. MapleStory by Nexon is often cited as an early example of this trend (Kong & O'Connor, 2009). By the second decade of the 21st century, the Free-to-Play model has expanded to mobile games on iOS and Android platforms where it truly blossomed.

Currently, it is by far the most prevalent monetization model on mobile platforms generating over 95% of the revenue. In the last decade, it has started to gain more and more traction on other platforms, PC especially but also console. The games such as League of Legends by Riot Games launched in 2009 and Team Fortress 2 by Valve are examples of early PC-based Free-to-Play games. Games such as Fortnite, Genshin Impact, and Apex Legends are some of the most popular titles across all platforms (ActivePlayer.io, 2023).

In the Free-to-Play model, a game is distributed for free via some digital distribution channel, for example, iOS AppStore, Android Google Play, Steam, or Epic Store, or even as a direct download from the publisher's website. Players are allowed and even encouraged to play the game for free.

However, the players are offered an opportunity to spend real money on various purchases within the game. The legacy name applied to those purchases, still encountered in many publications is *microtransactions* (MTX). The game feature dedicated to these transactions is called *Microtransaction Store*. This originated in the fact that initially these purchases were limited to very modest prices of a couple of dollars or euros. This term doesn't reflect the reality of the moment. As a game developer whose livelihood depends on such purchases, I do not want anything about them to be characterized as micro. The preferred term for such transactions is In-Application Purchases (IAPs).

Distributing a game for free has some obvious advantages. Above all, it greatly expands the install base of the game. No price, no matter how small can ever complete with something being free. This is clearly seen when comparing the download data for Free-to-Play with even the cheapest of premium mobile games, even when their price is $0.99 (Blacker, 2023).

Game developers still want and need to make a profit somehow. Since players are not obliged to pay to install the game, the objective becomes to motivate them to eventually spend some money within the game by making an IAP. They will do so only if they see the value in the stuff that the game is offering for purchase. The objective of the game design becomes providing and presenting value to the players and keeping them entertained so that would stick with the game long enough to make at least one or preferably more than one purchase within the game.

Thus all Free-to-Play games are run on the GaaS model. They are run as Live Services. Player *engagement* and long-term *retention* are the imperatives. This requires a fundamentally different approach to game design compared to the premium model. If we have compared designing premium games to writing a novel or making a feature film, designing a Free-to-Play game is more akin to making a long-running TV series. Most of this book deals with the design methodology focused on exactly these sorts of design tasks.

Notice that not all games that are run as a service need to also be Free-to-Play. Many games such as World of Warcraft are run as subscription-based services, while others are run on a hybrid model in which the players are required to pay the initial premium price but are subsequently offered to make purchases within the game, FIFA series by EA is a good example of this.

Both premium and Free-to-Play models can and will continue to coexist. The main reason is that not all types of games, even not all game genres are compatible with the Free-to-Play monetization model. As long as there are enough customers interested in purchasing such games, some gaming studios will continue to produce them.

On the other hand, the majority of people prefer Free-to-Play games. As I have already mentioned this is clearly evident in numbers. The number of downloads of Free-to-Play games simply dwarfs any install numbers for premium games. Companies keep making Free-to-Play games because people love them. The love of Free-to-Play games can be easily measured by several easy-to-quantify values, the number of downloads, the time users spend playing the game, and the amount of money some of them spend on the game. Gaming companies are after all run as business enterprises. Our aim is to entertain the world and numbers clearly show that players are being entertained. Otherwise, they would just drop our games and go do something else. With zero upfront price, there is no sunken cost to make them stay with a game that they do not like.

Free-to-Play games are fair to customers. The players get to play the game for free. They get to know the game before they can decide to spend money on it. In contrast to premium games, in Free-to-Play, you get to purchase things you already know you will like.

In most Free-to-Play games, you can achieve basically anything that you want without playing. In most cases, only a tiny fraction of players decided to actually spend money on a game. The metrics associated with this is known as *conversion rate* and represents the percentage of players that make at least one purchase within the game. Typically this is somewhere between 1 and 10% of players. Everyone else enjoys the game for free.

Teams developing Free-to-Play games need to respect free players because each and every one of them is a potential customer. If you run a business, you do not want to chase away a potential customer that has entered your shop!

Furthermore, you cannot make a successful business by exploiting your customers, at least not one successful in the long term. The world simply does not work this way. People are not stupid. If they don't get what they paid for, they will just go elsewhere. Unless you run a government monopoly, you can't go about screwing the people your paycheck depends on.

Free-to-Play players are some of the most sophisticated customers in gaming. They might not be following gaming news, nor care about the obscure lore of big gaming franchises. However, they are not naive, they are not conditioned and most certainly they are not addicts. People who spend on mobile Free-to-Play games usually know the games that they are playing inside and out. They understand the value of every item that they decide to purchase and will scrutinize the value of every new thing you offer them. They will take into account not only in terms of nominal value but also relative value within the very subjective context of their own playstyle. This is especially true for players who repeatedly spend money on the same game. Quite often they will budget their money and allocate an amount of money that they are planning to spend on their favorite game per week or per month.

Pricing Curve

Distributing games for free is one of the most noticeable aspects of the Free-to-Play business model. However, it is not the only important aspect. That zero price point ensures that the games will be installed by the broadest possible audience, by eliminating the initial purchase threshold. But that broad audience alone doesn't do anything to ensure the profitability of the game. This is where another important aspect of Free-to-Play comes in.

When it comes to premium games, each player of the game is expected to pay more or less the same fixed upfront price. It doesn't matter if the player stays with the game for hundreds of hours or if he drops it after mere minutes. The player already paid for the game, and that is it. This actually means that the most invested and the least invested players will end up paying the same amount for the game regardless if their needs and expectations have been satisfied or not.

On one hand, this is not really fair to the players that turn out not to be happy with the game, as they still need to pay. Conversely, this is not really good for the team behind the game as it puts a cap on the revenue that they might derive from their work even from their most dedicated and engaged players.

The Free-to-Play model solves this problem by introducing a *flexible price curve*. Nicholas Lovell describes this in detail in his book The Curve, which I can very warmly recommend (Lovell, 2013). To explain this concept, Nicholas Lovell, in his book, uses a concept taken from sports. How much does watching a game of football, the game in the US known as soccer, cost? Well, you can visit your local pub or a sports bar and enjoy watching the game for the price of a pint of beer. You can also watch it on your TV at home for the price of a cable TV or streaming subscription. A stadium entrance ticket for one game can be a bit more expensive. Teams usually offer a whole price range of various tickets. If you are a dedicated fan of a club you might buy a seasonal ticket. There are also tickets for VIP lounges and similar things. In addition to this, you might buy a team jersey or a scarf, or any of the additional team souvenirs. Each new monetization layer adds a new bit higher pricepoint and a new set of features available to purchasers. The price curve of football actually continues to astronomical proportions. Ultra-rich individuals are actually buying whole football clubs (Philippou, 2021)!

The metagame of football has been evolving for more than 150 years (Nixon, 2017). It had plenty of time to develop various very sophisticated monetization methods. Free-to-play games are doing essentially the same thing by providing multiple pricing points for various IAPs. In addition, they offer the opportunity to most engaged players to repeat their purchases over and over. In this way, there is no fixed upper limit on the amount of money that a player can spend on the game. On the other hand, since the initial price is zero, there is no lower limit on spending required from the player.

If you are even casually familiar with the Free-to-Play model you might have heard some of the terms by which various types of players are described depending on the amount of money they spend within the game. As I mentioned already, the vast majority of players play these games without paying anything, they are usually called *free players*. The remaining, i.e. paying players, are often divided into three categories, the names of which are all derived from various types of sea creatures:

1. *minnows*—the ones that make a single or very few very small purchases,
2. *dolphins*—that spend moderate amounts of money,
3. *whales*—the players that are known to spend relatively big amounts of money within the game.

Personally, I am not particularly fond of these terms and prefer to use the term *superfans*, popularised by Nicholas Lovell, for our best customers. These categories are usually defined in terms of the total amount of money that a player has spent within the game. The boundaries between these categories are arbitrary and differ significantly from game to game. In general, this nomenclature, and indeed the way of thinking is already some-what outdated. When building a game economy, it is better to focus on providing value to all the players. The most valuable customers are often not the ones that make the biggest individual purchases, but rather the ones that keep regularly and steadily spending the many within the game. To these people, a game becomes a true digital hobby.

Why Do People Buy Stuff?

Back in 2012, when I was just starting my game design career, Free-to-Play games were, at least in the West, a brand new concept. Quite a lot of people were still trying to come to grips with the notion that selling virtual items could all of a sudden be big business. The question that everyone was asking was why would anyone in their right mind spend money on virtual items. The company that I worked for at that time was good enough to pay for and organize a workshop with Nickolas Lovell. I remember him pointing us to an even older Wall Street Journal article on this topic (Constable, 2009; Hamari & Keronen, 2017; Liew, 2009).

It is remarkable to see that things didn't fundamentally change since then. The reason for this is that motivations driving people to spend on virtual items are deeply rooted in human psychology.

So, why then people spend on virtual items?

They spend on virtual items for exactly the same reasons they spend on physical real-world things! In this text, I'll reiterate what these motivators are and how they are catered for in current Free-to-Play games.

The purchasing motivators may be varied, but they all belong to three broad categories. People buy stuff to:

- Be able to do more,
- Establish identity,
- Establish and maintain relationships.

Doing More—Efficiency, Tools, and Access

Obviously, we spend money to buy stuff that is useful. Things are useful if they allow us to do more than we could without them. Things that we buy allow us to overcome the limitations of our physical bodies or the scarcity of resources we are facing at any given moment.

If I am hungry, I am facing an acute scarcity of eatable items in my hands. To solve this, I can buy a sandwich. Therefore, by spending money we can obtain things that:

- Can immediately satisfy our physical needs: food, drinks, clothing, paying the rent, etc.,
- Help us obtain other things, i.e. tools that allow us to make things that we need or want,
- Increase our efficiency in doing tasks.—Typing on a computer is more efficient than on a typewriter, which is, in turn, more efficient than handwriting; buying tram tickets increases my efficiency traveling to work every day.
- Grant us access to other things or places—a cinema ticket!

Spending money to get the ability to achieve more is the cornerstone of monetization in Free-to-Play games. It was the first class of motivations that were integrated into these types of games and is consequently the most developed one. I will mention just some of the examples.

Buying efficiency in many games takes the form of purchasing the missing resources such as coins, gems, card packs, etc. A common design trope is that two resources are needed to make an upgrade of some element in the game. For example, upgrading a Character in Clash Royale requires Gold and corresponding Character Cards. These games create an artificial scarcity of certain resources. The game's virtual economy is tuned in such a way that one resource is overabundant, but the other one is relatively scarce. Typically, in the late game, the player will have a lot of Characters Cards that can be upgraded but would be chronically short on Gold. Making a purchase of gold coins reduces the perceived inefficiency.

Buying power-ups in casual games is another way of purchasing efficacy. Quite often a player gets stuck for a long time on a particularly hard level in Candy Crush (Fig. 2.1). Buying a power-up in the hope that its use could help him get out of the uncomfortable situation, is an example of this behavior.

Buying more energy in casual games with classical energy mechanics is another example of buying efficiency. The player is forced to stop the inherently pleasurable activity and wait for the energy meter to fill up. The player will part with money to skip the waiting.

Fixing time planning errors is yet another common example. This is the key to the success of many farming/crafting games, such as HeyDay or Township. Puzzle in this type of game revolves around the optimization of production queues of the virtual farm or two. Get weed from the field, get apples from the orchard, produce animal feed, produce flour, give food to cows, get milk from crows, etc., etc. All these activities involve waiting. Success requires time planning. Planning errors are time-consuming to fix. The player pays to fix his own mistakes.

Buying access in games usually revolves around paying for the game itself or at least for a part of it. The classical premium model is built on this. Players need to pay a certain amount of money in order to get access to the game itself. Paywalls, where the players

Fig. 2.1 Powerups in Candy Crush Saga

are expected to pay for a portion of a game, are another example (Fig. 2.2). Ultimately all subscription services work in this way.

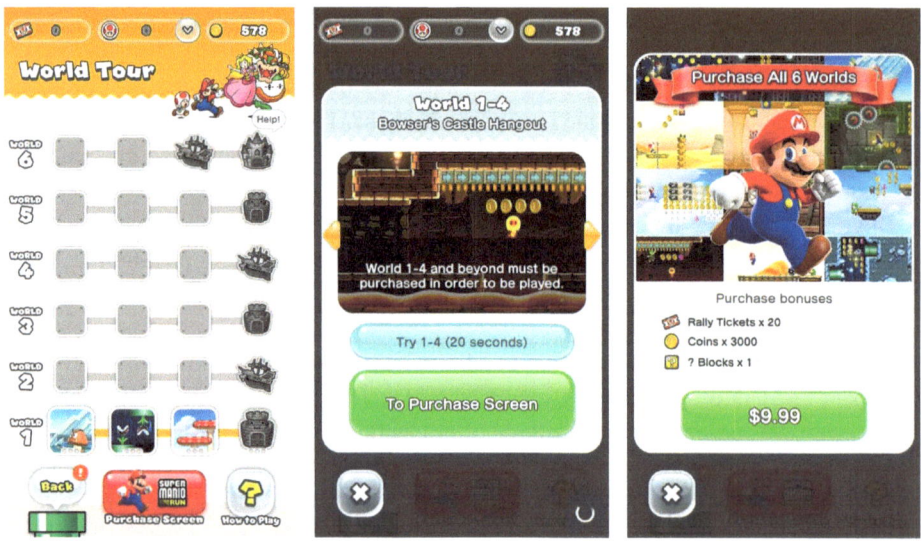

Fig. 2.2 Super Mario Paywall on iOS

Establishing Identity

People buy stuff to express their identity. Establishing our own identity as distinct individuals is fundamental in our psychology. In real life, buying clothes is only somewhat related to keeping our bodies warm and protected from the elements. There is so much more about our social identity interwoven into the choices that we make when selecting clothing.

There are actually several distinct motivators at play here. A desire to stand out is rooted deeply in the process of sexual selection. The best way to show that you are possibly a fit sexual partner is to show how much energy and resources you have at your disposal (Zahavi, 1975). The best way to show this is to waste them on superfluous things. This behavior is a part of the evolution of organisms on this planet. It is responsible for the evolution of things like peacock tail feathers for example. It is a very big factor in human behavior as well. We like to show off our wealth the same way the peacock likes to show off its tail.

The opposite of this is our desire to fit in. It is evolutionarily younger than peacocking but still ancient. Humans are social creatures. We are meant to work in small tribal groups. An appearance is a form of establishment of group identity. Wearing an article of clothing signals to others something about our identity. Real-life examples include buying a brand t-shirt or a football club scarf. It fulfills our desire to belong to a social group of similar and like-minded individuals.

Fitting in to stand out is a combination of the previous two notions. Group identity is further strengthened by this behavior. The cohesion of the group members is reinforced by the uniformity of appearance. Distinct group appearance helps group members to stand out from the other groups. Real-world examples include buying clothing that belongs to any of the particular subcultural groups. A Hipster is recognized in the crowd the same way a Goth was recognizable in the early 2000s or a Punk in the 1980s.

Typical examples in gaming involve so-called Vanity Items i.e., items that have aesthetic value only. See, for example, hats and pets in Among Us, costumes and skins in games such as Fortnite and Brawl Stars, etc. (Fig. 2.3). These elements have contextual value. Context is formed by the audience. In order for these types of items to make sense, the game needs to provide a human audience, i.e., it needs a strong social element.

Maintaining Relationships

Some 75,000 years ago, a volcano went off on the island of Sumatra. The eruption was so severe that it caused a temporary climate change, which in turn created massive dying off of many sorts of animals. Humans nearly went extinct. Our population declined to less than 30,000, a population of a small town. The evidence for this is seen in our DNA (Ambrose, 1998). The only humans that evolved were the ones that were parts of tribal

Fig. 2.3 Character skins in Brawl Stars

groups that learned how to collaborate with other similar groups. Contrary to popular belief, altruism is the key to survival in harsh times, and selfishness is something that can proliferate only in times of plenty.

In order to survive, humans had to collaborate. In order to collaborate, they had to establish a certain set of rituals. Gift-giving is a cornerstone of this behavior. Giving a gift signals one's ability to overcome selfish interests and build a relationship with the other person or a group. In an unwritten protocol that we inherited from our ancestors, the act of gift-giving demands reciprocity (Yunxiang, 2020). It is a powerful tool that binds society together and allows the formation of new relationships.

In real life, in almost every culture in the world, people spend enormous amounts of money on gifts that they send to others. We have instituted whole seasons that revolve around this activity. We have Christmas, St. Valentine's, Ramadan, and the 8th of March.

The concept of gift-giving as a monetization driver is somewhat underdeveloped in video games. I am not talking about sending virtual gifts that the player can obtain by simply playing the games. These things have been present in the Free-to-Play world since the early days and are still present in some games. See for example sending gifts in Pokémon GO!

What I have in mind are examples where a player is invited to spend either actual real money or at least hard currency in order to be able to send a gift to another player.

Probably the best example comes from Fortnite (Fairfax, 2020). Here players can spend money and send Battle Pass as a gift to another player (Fig. 2.4). The hope here is that the receiver of the gift would feel obliged to return the favor and would perhaps be

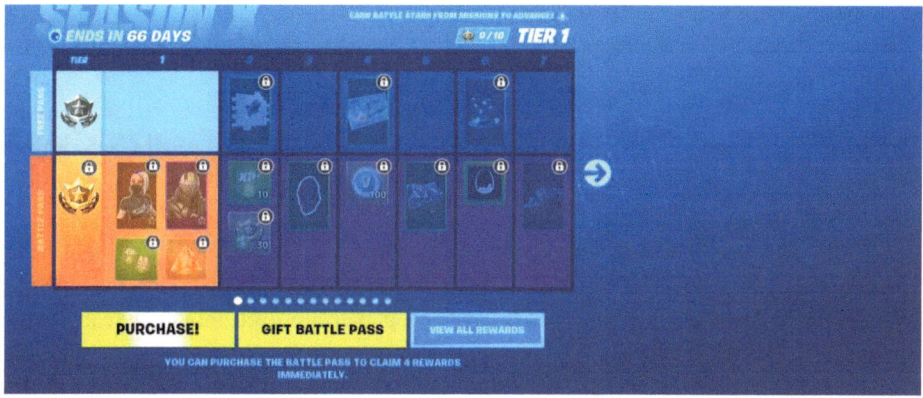

Fig. 2.4 "Gift Battle Pass" option in Fortnite

motivated to make a similar purchase. Ideally, this would create the virality of conversion of non-paying players into payers.

I believe that this type of motivation has the biggest unexplored monetization potential. It will be interesting to see how this concept will evolve.

Quantitative Metrics

Operating a game as a service offers several important advantages over publishing it as packaged goods. Above all the game can evolve. Arguably a game run as a service is in its worst shape at the moment when it is launched.

The software architecture in which this type of game runs consists of two distinct components, a client application running on the player's device, a smartphone, a console, or a gaming PC, and the server backend. The client and the server are in constant communication. This allows the team running the game to constantly collect data about player behavior within the game. The technical term for data gathered by the server from the client is *telemetry data.*

Contrary to popular opinion this data is not player focused, but rather game feature focused. We are not interested in how specific individual players interact with the game, rather we are interested in how large groups of players interact with the specific game features. This is an important distinction. Individual data of a specific user does not hold a lot of value in isolation. This is what invalidates most of the sell-your-own-data business propositions touted by Web3 proponents (Szyller, 2022).

The telemetry data is used to derive a series of quantitative metrics known as *Key Performance Indicators* (KPIs) which are used to determine the health of the service from the business perspective. These can metrics can be divided into two groups:

- universal metrics that apply to all games run as a service,
- game-specific, or even-feature specific metrics that are important in the context of an individual game.

I will mention the most important universal KPIs as they are common to all GaaS projects. The other group of metrics by definition differs greatly from game to game and is tailored by individual game teams for their specific needs.

These metrics apply to any application run as a service. They were, indeed, created to measure the performance of applications and not games specifically. This is sometimes reflected in the terminology. The words user and player are used interchangeably in the following text.

General KPIs in turn include two closely integrated categories:

- retention-related KPIs, and
- monetization KPIs.

Monetization-related KPIs are a direct indicator of how well the game is operating as a business. The game is distributed for free and its revenue is derived from the purchases that the players make within the game, i.e. IAPs. Since a lot of players are able to play the game for free, it is obvious that the game team needs to know what percentage of players are actually spending any money on the game. This percentage is usually called the *conversion rate*. The word conversion here comes from the idea that a non-paying player converted into a paying player by making at least one purchase within the game.

As we mentioned above the Free-to-Play model does not put a limit on the maximal amount of money a player can spend within the game. Therefore, it is good to know how much money players on average spend within the game, the *Average Revenue Per User* (ARPU), and *Average Revenue Per Spender* (ARPS), also known as *Average Revenue Per Paying User* (ARPPU) provide this information. These numbers are derived by deriving the game revenue by the number of all players and the number of paying players respectively.

Finally, since various players spend very different amounts of money within the game, it makes sense to calculate how much money in total individual users spend within the game during the total time that they played the game. This is known as the *Lifetime Value* (LTV) of a user. This number is a sum of all purchases made by the individual user and is usually calculated as an average for the whole user base or for specific subgroups of users.

These metrics are calculated as averages per user. The total revenue of the game depends on the number of active users. This is typically expressed as the number of Daily Active Users (DAU) or Monthly Active Users (MAU). The definition of the active user can differ. Most often, any user who started the client application, i.e. logged in, during the specified period of time, is labeled as active.

One curious and important thing to keep in mind regarding these quantitative metrics in the digital world has to do with their distribution. Variables in the physical world quite often exhibit the Gaussian distribution. Variables in the digital world most generally follow Poisson distribution (Haight, 1967).

The metrics that I just listed deal with the revenue of the game. In order to calculate the actual profits we need to take a look at the other end of the equation, i.e. to cost of getting the same people to play the game in the first place. The key indicator of this is *Cost Per Install* (CPI), which is defined as the total amount of money spent on advertising the game, i.e. on *User Acquisition* (UA) divided by the number of players that have installed the game. This in turn can be used to calculate the Return on Ad Spend (ROAS), i.e. the revenue by the amount of money spent on the user acquisition.

In the Free-to-Play mode, the players are not required to pay any upfront price for playing the game. The hope is that they would eventually convert from non-paying to paying users of the game. In order to maximize the probability of this happening it is important to ensure that they stay playing the game as long as possible. Furthermore, paying players often make several purchases over a period of time. Some of them are even doing this regularly. Keeping players from quitting the game is what is known as player *retention*.

There are several metrics used to measure this. Above all, there is a *retention rate* measured after a specific period of time. It shows the percentage of players that continue to play the game after a specific number of days since installing the game. It is usually calculated at least after one day, after one week, and after one month of gameplay, with intermediate calculations on day three and day 14 also being common. These are usually abbreviated as D1, D3, D7, D14, D30 etc. The inverse of the retention rate is the *churn rate*. The word churn indicates the act of abandoning a game.

The *first-time user experience* (FTUE) is a somewhat fuzzy term denoting usually the first interaction that the player can have with the game after installing it. It is most often applied to the first play session that a new player has with the game. It can include a specific tutorial but it can also go beyond both the tutorial and the first play session. We are going to discuss the intricacies of designing FTUEs in Chap. 10 when we discuss practical game design examples. In general, FTUEs tend to be handcrafted and tightly controlled player experiences, consisting of a series of distinct steps. Usually, the *churn rate* is calculated for each of these steps separately. The collection of churn rates of all FTUE steps is referred to as the *FTUE funnel*.

Finally, since the GaaS is run over a long period of time, with new players constantly coming in and the old players constantly flowing out of the game, all these quantitative metrics are expected to drastically evolve over time. They are periodically recalculated and updated for the user base in general. However, they are also routinely calculated for the subsets of users defined in terms of the period of time when they installed the game. These subsets are referred to as *cohorts*.

Subscription Model

It started with Netflix. It's the year 2021 and subscription services are everywhere. Apple is hawking its Arcade, Google is trying to convince you that you need more storage on your Drive. Microsoft desperately wants you to use its SharePoint. I am pretty sure that soon the likes of Roomba or Neato will in the near future launch their own premium subscription service and that you won't be able to use your robot vacuum cleaner without paying $10 per month.

The same trend is evident in the world of gaming. There is already Xbox Live Gold. Apple has been pushing heavily its Apple Arcade, first as a standalone service, and now as a part of Apple One, a bundle of its other cloud-based services. They have been also aggressively promoting the idea of subscriptions to game development teams running Free-to-Play games on iOS, by offering better revenue-sharing terms.

Limits of the Model

To be sure, subscription services have their advantages. To some players, they are seen as the more appealing option than Free-to-Play or a premium model. To some of the companies, they represent a more predictable business model offering seemingly a more stable source of revenue.

However, subscriptions as a business model are nothing new. Newspaper, magazine, and cable TV industries have been employing it for decades. As they have learned, this business model has one fundamental flaw that severely limits its growth potential. This is the lesson that these industries have learned the hard way.

If you are running a game as a service, it pays off to at least be aware of it. This is especially important if you have experience in operating a successful Free-to-Play game.

The subscription model puts a hard ceiling on the amount of money users can pay for your service. This should be obvious. Assume that you are charging a subscription of $4.99 per month. The revenue that you can expect from every user that stays with your service for a full year will be 12 x $4.99 = $59.88. This is a decent amount of money and as long as the cost of acquiring your users and the cost of production of the content that they consume remains below this level, you should theoretically be able to profitably run your business.

The interesting things, however, begin to happen once a company starts approaching the practical limits of its user base. This is the predicament in which Netflix and HBO are getting into right about now. By now pretty much everyone that cares to have one already has a Netflix account. As the user revenue has a hard limit, the revenue growth begins to stall. Retaining the existing users becomes the priority.

This is the point at which the true limitations of the subscription model start to unfold in several ways. In these circumstances, a company has two prime objectives:

- retaining the existing users,
- growing revenue.

There are essentially two ways in which subscription-based businesses can retain their users. The first one is a dirty little secret of this revenue model.

There is certain inertia built into the core of the subscription business model. Companies love subscription services because they seem to be providing a more stable source of revenue than any monetization model that is based on occasional purchases. Simply put, the company charges each user a fixed amount of money each month regardless of whether he consumes the content or even engages with the service.

This of course hinges on an assumption that users once subscribed will stay subscribed in perpetuity regardless of whether they actually use the service or not. This is a fair assumption. Most people will not bother to cancel a service that they are not using very much. Opting out of the service you have been subscribed to requires overcoming natural mental inertia. It is similar to overcoming the natural inertia of making an occasional purchase. However, in the former case, inertia works to the company's advantage, in the latter against it.

This works only if the cumulative price that the user pays per month for this and additional similar services is low enough. If the total monthly spending rises over a certain threshold, players seem to change their behavior. Savvy users are switching between Netflix, HBO, etc., turning their subscriptions on and off on a monthly basis, and binge-watching the content.

Companies are well aware of this problem and are developing strategies to combat it. HBO is already offering a 50% lifetime discount that is valid only if the user doesn't interrupt his subscription ever. This is already a hefty discount. I don't have insight into their user behavior data, but I am sure that they are willing to roll with it because numbers make sense to them. This would imply that an average user is subscribed to HBO for less than six months in a year. Otherwise, this strategy of user retention will start to work against the company's other objective, the growth of revenue.

The other way of retaining users goes back to the core of any business. It is about providing value for your customers. Content is king. From the user's perspective, the best way to keep them is by providing value for their money. In practical terms, this means a greater amount and better quality of the content that you offer. This applies to TV shows, movies, music, or games. Of course, this will inevitably lead to an increase in the company's production cost. Inevitably as competition grows more fierce, the companies will be spending more and more money to produce new content. According to some reports, Netflix alone will spend $17 Billion on production in the fiscal year 2021 (Carcasole, 2021).

Again obviously this goes against the other objective, i.e., the revenue growth, as these expenses start eating into profit margins.

Building a value proposition for potential users is also a part of revenue growth. You will attract new users more quickly if they perceive the value of your offer.

If a company is also an IP holder and producer of content, this task might seem easy. However, this can be a problem for a company that makes a transition from other monetization models to subscriptions. In order for a subscription to work, the company needs to make available its complete portfolio (especially its flagship IPs) for a fixed monthly price.

This is a problem if the revenue generated by individual items in the portfolio is greater than what would be generated via subscription over the same period. Consider a gaming company that makes a profit of 50 dollars per average player per year for each of games A, B, C, and D. In total this makes up to 200 dollars. If the said company switches to the subscription model and makes games A, B, C, and D available to all users for a fixed price of 10 dollars per month this amounts to only 120 dollars per year. In other words, the company would be making 80 dollars per year less than it would otherwise. Here lies a Catch-22 of this model. In order for a subscription to be attractive to users, it should offer more value than the purchase of individual content.

This is the problem that established traditional big names in the gaming business, such as EA, Ubisoft, Bethesda, etc., are facing when it comes to subscription services.

Finally, a company can always try to grow its revenue by hiking up the price of its service. This is a valid approach. However, this doesn't allow the company to break away from the inherent limitations of the business model. The revenue ceiling will simply be pushed to another level.

Hiking up the prices goes directly against the building of the value proposition. It can thus be detrimental to user retention. In order to offset this, the company might resort to offering larger more, and better content, in turn increasing the production cost and eating into the profit margin. Another Catch-22!

Some companies have been experimenting with augmenting their subscription revenue with ad-based income. Ad-based business doesn't suffer from the same fundamental limiting factors. There is no upper limit for the amount of money a company can charge for an ad placed in a strategic place. See for example astronomic sums spent on Super Bowl ads.

Adding advertisements to your content again goes directly against the value proposition of many of these companies. This is especially relevant to gaming since the subscription has been offered to players specifically as a way of avoiding ads and IAP monetization.

On the other hand, ads have been a part of the magazine and cable TV business model for decades and haven't been able to save them from inevitable doom.

Navigating this conundrum is what makes the subscription business model so difficult. Some companies such as Netflix have been able to so far escape this trap, although the speed by which they have been adding new users has been slowing since early 2020, see Fig. 2.5. Other attempts have already run into headwinds. This is the reason why Apple is bundling Arcade with other services.

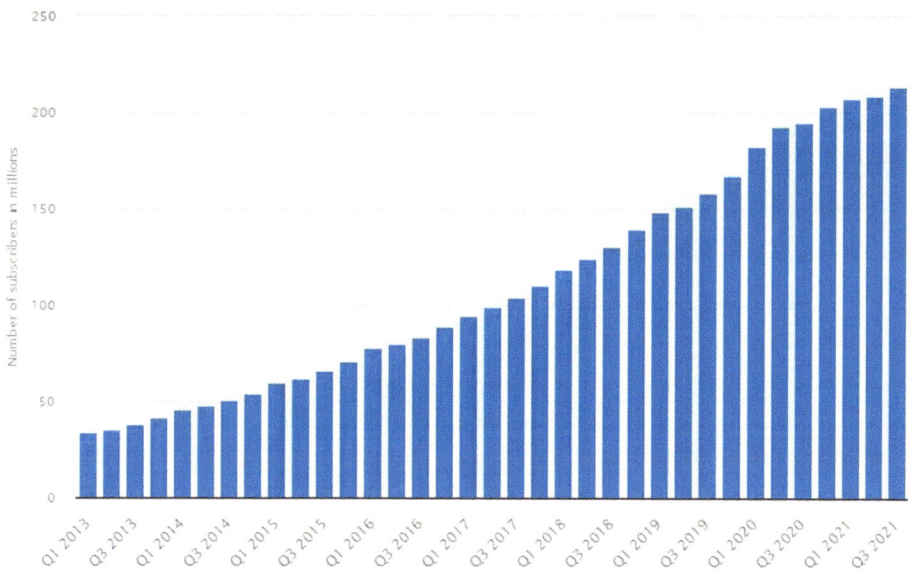

Fig. 2.5 The number of Netflix subscribers per quarter from 2013 to 2021, according to Statista (Stoll, 2023)

On the other hand, the data shows that subscription-based companies are still growing and that it is actually growing faster than companies relying on other monetization models (Neele & Speetjens, 2022). It is going to be interesting to see how the market will evolve.

Video Ad Monetization

Imagine that you are walking into a supermarket. You want to buy toothpaste and perhaps some snacks. You walk down the aisle past the bags of chips and 300 kinds of breakfast cereals. Finally, you spot what you are looking for. You reach for it ready to pick it off the shelf and put it into your shopping basket, but at that very moment, a salesperson pops up seemingly out of nowhere. She grabs you by your arm and hands you a leaflet advertising the competing store from access to the street. The salesperson doesn't let go of you until you have read everything that is printed on the leaflet. She doesn't even let you move your eyes away from the glossy piece of paper in your hand. You are baffled by the experience but you have no choice but to comply. Things are different in the digital world, but this is essentially what happens when a game shows you a video ad.

Monetizing Free-to-Play video games using video ads is an increasing trend at the moment. Many people advocate it as a solution to all the perceived ills of more traditional methods of generating revenue from games distributed for free, such as in-app purchases

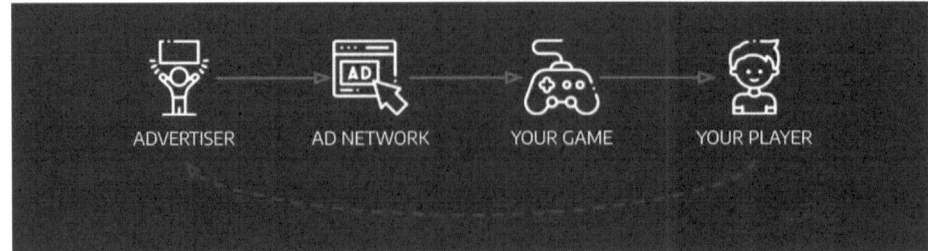

Fig. 2.6 A schema of a typical Ad system

of virtual items. At first glance, this proposition might seem true. The player is never required to spend any actual money in the game and is often rewarded for his time with at least some virtual item.

However, at a close look, one can quickly discover that this method of monetization is built on top of conflicting motives of all involved parties. If you are employing this monetization model in your game, it pays off to study this conflict in order to better prepare your own business strategy.

There are no less than four sides involved in each implementation of this mode, as shown in Fig. 2.6:

- Players—the audience for the ad,
- Game Creators—the studio that runs the game and sells advertising space,
- Ad Providers—a company that mediates between the Game Creators and potential Advertisers,
- Advertisers—a company that provides the ad and pays for the ad to be shown.

All of these four parties are guided by distinct interests that are in many ways at odds with each other. We should examine them further in detail, by taking a look at the most important interconnections in this framework.

Game Creators and Players

Let's begin this discussion from arguably the key connection, the connection between Game Creators and their audience, the players. Without this connection, all others would be obviously irrelevant.

As a player, you might be forgiven for thinking that you are a customer of a company/ studio that published the game. After all, you are actually the one using their product. If the game offers any other monetization method, and if you have chosen to spend money on it, this is very much true. However, in the context of video ad-based monetization, the product that the company sells is actually yourself!

In this business transaction, the video game company is selling your viewing time and attention to an advertiser company. You can rest assured that both companies are well aware of this and will be treating you accordingly. Your satisfaction with UX will be viewed primarily in terms of increasing the likelihood of you actually choosing to engage with offered video ads.

At the moment, most of the games seem to follow the model where the player is given a choice to interact with ads and is incentivized to do so using rewards that at least have contextual value within the game. Ideally, the players at least feel that trade-off of their attention and viewing time for a virtual item is a fair one. The situation is different in the case of games that remove this choice from the players and force-feed the player a steady stream of interstitial ads.

In some genres of games, video ad monetization is employed as a model complementary to other monetization mechanisms. In this model, Game Creators are supposed to give out virtual items to players as compensation for their time and effort, as these items are supposed to feel contextually valuable to players. These are items that Game Creators are also likely to sell through their other monetization mechanism.

This creates another conflict of interest, as a player you are motivated to seek for free the items you would otherwise be expected to buy. From the Game Creator's perspective, video ads as a monetization mechanism entail the potential of cannibalization of other models of monetization.

Game Creators have a vested interest in giving you the least valuable virtual reward, thus minimizing the potential damage to their other monetization models. In turn, they are going to try to devalue the player's engagement time!

Game Creators and Ad Providers

The next in the chain is the connection between the Game Creators and Ad Providers, i.e. a company that acts as a middleman between the publisher of the game selling the ad space and the potential advertisers.

Engagement, i.e., the time that the player spends within the game is one of the key performance indicators of every Free-to-Play game. Game Creators invest every effort to maximize it. In simple terms, as a Game Creator, you want your users to stay playing your game as long as possible.

Furthermore, the Game Creators are actually in most cases paying substantial sums of money to attract players to their own games. User acquisition is the single biggest expense that a studio running a live game will encounter.

Ad providers, on the other hand, get paid for every new install of the game that is being advertised. After all, they are on the receiving end of the user acquisition transaction.

Therefore, they would actually like to have the very same player leave the game that served them the ad and install a totally new game that has just been advertised to them. The conflict of interest is obvious!

Simply put, as Game Creators we will try to sell Ad Provider's product (an ad) but we at the same time hope your product fails in its only purpose (motivating the player to abandon the current game and switch to the advertised one).

The problem is exacerbated further by the way that most current implementations of the system work. The Game Creators are responsible for the part of the UI that is related to the very beginning and the very end of the video ad flow, i.e., the point in the game where an ad is offered to the player and to the point in the game at which the player is supposed to claim his reward for watching the ad. The Ad Providers are responsible for the part of the flow in between, including interactive UI elements of the video ad itself. We have two companies trying to optimize two parts of the UI flow for exactly the opposite goals, wanting at the same time the player to leave the game and stay within the game.

Furthermore, Customer Support is almost always the responsibility of the Game Creators. The ad-providing company will try to tweak the UI flow of the ad in order to maximize the click-through rate. They are not concerned with the quality of the player's user experience. The Ad Providers are happy to leave the game company to deal with the potential consequences.

If your game has an average play session of around 5 min and you are serving around 3 to 4 ads during this time, you are essentially outsourcing the UX of 30% to 40% percent of the time your player is spending within your game to an external company!

Ad Providers and Advertisers

Also curious is the relationship between the Ad Providers and their customers, the Advertisers. At least in broad terms, the interests of these two parties align. The Advertiser would like to get as many new players as a result of an ad being shown. The Advertiser pays according to the effectiveness of these ads.

The Ad Providers are paid based on the click-through rate of the ads that they serve. They have control over the UI flow of an ad, i.e., the functionality of the interactive elements of the ad. Non-interactive creative assets, including visuals and audio, are on the other hand most often provided by Advertisers themselves. Thus, the Ad Provider company's revenue depends on something that they have only very partial control over. Bad or unattractive visuals are far more likely to be a cause of a bad click-through rate than the unoptimized UI flow. Advertisers are thus expecting an Ad Company to maximize the effectiveness of visuals that they themselves provide to that company.

Simply put the Ad Provider is expected to turn into gold whatever material is handed by the Advertisers. In this setup, the tool that an Ad Provider has at its disposal to maximize its profits is to manipulate the behavior of ad viewers in order to maximize the click-through rate.

Ad Providers and Players

Finally, the loop closes with the relationship between players, as the Ad Audience and the Ad Providers.

In this arrangement, Ad Providers and the player are connected indirectly. Ad Providers do not have an incentive to care about players. What they care about is the maximization of the click-through rate. They are on the other hand in a sandwich between the Game Creators and advertisers.

They have very little reason to care about players' interests. Instead, they focus on maximizing certain player behavior. Players are a product that Ad Providers are buying from the Game Creators. As a player, you should fully expect that you will be treated as a product by this company. They will have no qualms in trying to manipulate you or even outright try to trick you into doing a certain set of actions.

On the other hand, the players are fully aware of this. They are rarely motivated by curiosity about what ads actually have to offer. Most often, the only thing they care about is the reward with which the Game Creators motivate them to watch the ad. The actual ad with its creative content is a nuisance for them, and they will do all that can to avoid watching it and yet claim the reward.

They will try to skip it, or alternatively, they will start an ad and then switch their attention to the TV or another screen for the thirty seconds it takes the ad to play. They will go as far as to train their cat to tap on ads, and do all sorts of other ridiculous things just to avoid using the product in the way it is intended!

Advertisers and Players

The final relationship that we will examine is between the Advertisers and the Players. The advertisers are of course paying for the ad to be shown to players in order to attract them to their own game. Players play games for various reasons, but mainly to occupy moments of their free time. At the moment when an ad is shown, a player is already playing another game. Chances are that he is already heavily emotionally invested in this previous game. At the very least, it is reasonable to assume that the player is somewhat entertained by the game he is playing. Therefore, the player has some inertia toward switching his attention to another game.

It is this inertia that the Advertisers are trying to overcome. The inertia is also one of the factors influencing the cost of user acquisition. The Advertisers will do their best to

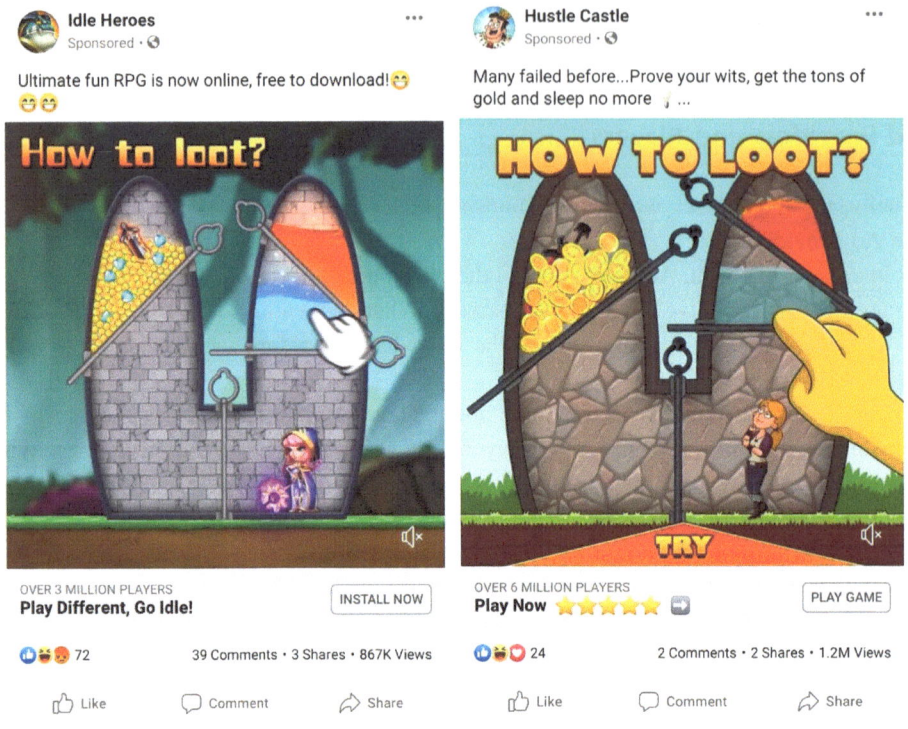

Fig. 2.7 An example of a misleading ad

maximize the number of players their ads attract. Over the last few years, they have managed to formulate a data-driven approach to this problem. The basic strategy employed by the Advertiser is to create several versions of the creative content for any given ad and run all of these versions in parallel to disjoint sets of viewers in a form of an A/B test. The best-performing version is then shown to a large audience.

The major breakthrough in this area came when some clever Advertisers realized that the best-performing ad content does not have to have anything with the actual game that is being advertised. As a player, you have no doubt already faced such misleading ads! Example of such misleading ad is shown on Fig. 2.7.

The Way Forward

As we have seen, video ad monetization involves several parties with different and sometimes opposing motives. To be absolutely fair, despite these inherent conflicts, ad monetization is a healthy thriving business, with all participants being at least somewhat satisfied. Understanding their motives is a solid starting point from which we could make things better. I firmly believe that being aware of the interests of all parties can lead to improvements on every level.

As a Game Creator, for you, the best way of navigating this conundrum is to respect the interests of all parties involved but to constantly protect your own interests. Pay attention to the needs of your players, provide them with incentives to watch the ads, but also try to provide them with incentives to return to your game, even if they have moved out to explore the thing that was advertised to them.

Try to work with as many Ad Providers as possible. Compare the experience you have with each of them and balance the need to maximize the short-term ad revenue vs. the long-term impact on your player retention. Do not neglect to provide feedback about the quality of service and ads being served by the Ad Providers, even if it seems futile. Be ready to ditch any Ad Provider that fails to respect your game and your players.

As an Advertiser, respect your future players. Do not try to mislead them with ads. Rather focus on bringing to light the best qualities of your game.

Player Psychology

3

Psychological Needs

Why do people play games? Well, that's easy… Games are fun!

I still clearly remember one lunch break from a couple of years ago that I spent going for a Pokémon GO raid. It was wintertime in downtown Helsinki, Finland. The weather that day was −7 °C (that is 19F in the US) with gusts of strong wind and snowfall. I was sure that I'd be alone, but as the scheduled start of the raid was approaching, a small throng of people started forming. By the time the raid started, there were about twenty of us, all middle-aged office workers from nearby buildings, spending our lunchtime chasing virtual monsters in the middle of a blizzard. In the end, I didn't even catch that legendary Pokémon I was after, but I had fun!

People do lots of stuff because they are fun. Some people think that it's fun to shoot clay pigeons. Others get a kick out of collecting ancient postage stamps. Some people spend nights bent over arcane rulebooks rolling twenty-sided dice. To millions, playing video games means milking virtual cows to produce virtual cheese and donuts in Township. Others scoff at such pedestrian trivialities and instead opt to play-act their fantasies of being a Witcher.

Well, what do games as diverse as Overwatch, Tetris, Township, or Among Us actually have in common? What makes all of these diverse bits of software successful games? …and what is fun anyway?

The definition of fun is a slippery one. It falls into that weird category of concepts that are easy to recognize but hard to express with words. There is a considerable body of work published on this topic. What I find helpful is to break this troublesome notion into more manageable parts. Something that we can define more precisely and put into words more easily.

© The Author(s), under exclusive license to Springer Nature Switzerland AG 2024   37
S. Stanković, *Game Design for Free-to-Play Live Service*, Synthesis Lectures on Image, Video, and Multimedia Processing, https://doi.org/10.1007/978-3-031-56156-6_3

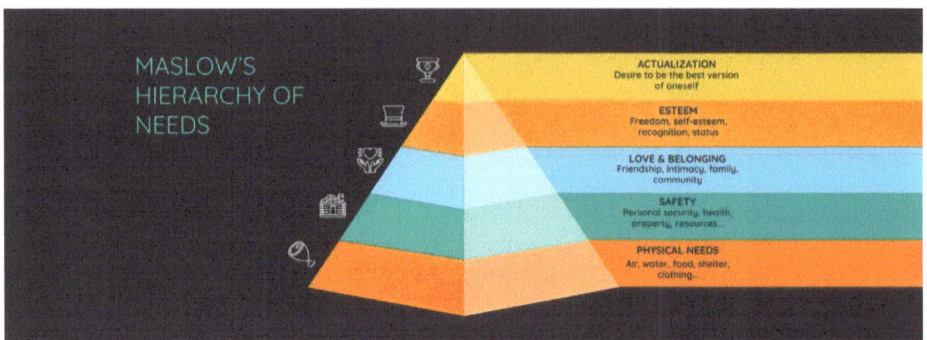

Fig. 3.1 Maslow's hierarchy of needs

Enter Self-Determination Theory (SDT), a psychological framework that can help us in this undertaking (Ryan & Deci, 2000). This theory has been originally developed by Edward L. Deci and Richard Ryan in the 1980s. Since then, it has been applied to diverse areas of human activity, including wellness and well-being, sports performance, workplace performance, and job satisfaction, to name but a few.

It was applied to the study of games by Scott Rigby and Richard Ryan via their The Player Experience of Need Satisfaction (PENS) Model (Rigby & Ryan, 2004).

The Self-Determination Theory speaks about human needs, building on the idea of the hierarchy of human needs proposed by Maslow in his Self-Actualization Theory (Maslow, 1943) (Fig. 3.1). It postulates that all humans share a set of at least three basic psychological needs:

- Autonomy,
- Mastery,
- Relatedness.

All humans seek out the satisfaction of those needs in the same way we require air, water, food, or shelter to satisfy our basic physiological needs. Humans will naturally gravitate towards activities that offer more chances for the satisfaction of their needs. There is a sizable body of psychological research that confirms that this behavior is universal regardless of the age, gender, or culture to which any particular person might belong.

Obviously, some human activities are better at satisfying these than others. An hour spent stuck in a traffic jam is not really doing wonders for either our sense of autonomy or competence; quite the contrary. On the other hand, video games are actually designed in such a way to offer psychological need satisfaction in spades.

So what do you do when you are stuck in that traffic jam? Well, of course, you reach for your cell phone and start chatting with friends to fill in your acute shortage of Relatedness or start playing a game to do the same for Autonomy and Competence.

This is what fun means. It's an activity that helps us to satisfy the three basic psychological needs! In other words, people play games because they offer a great degree of need satisfaction.

A well-designed game will afford the satisfaction of one of the three basic needs to an extremely high degree or at least two of those to a great degree. Truly successful games will offer the satisfaction of all three needs to a very great degree.

To fully understand what this actually means, we should take a look at more precise definitions of these three needs as seen in the context of Self-Determination Theory.

Autonomy

As a term, Autonomy is instantly recognizable. In everyday speech, it is often used as a synonym for words such as freedom, liberty, or independence. However, the SDT uses a very specific and relatively narrow definition of autonomy:

Autonomy is a desire to be a causal agent of one's own life and act in harmony with one's integrated self. This narrower definition of Autonomy actually excludes quite a few common associations. Specifically, in this context, Autonomy is not freedom in a general sense. Freedom is the absence of constraints. In the SDT framework, autonomy means that we act with awareness of external constraints. Furthermore, Autonomy does not imply a lack of demands. Rather, it is a response to the demands. It does not mean individualism. Autonomy, as a need, is evident even in cultures that are not individualistic in nature. Finally, Autonomy is not equal to independence in the sense that one acts alone. We can act autonomously and still rely on the help and guidance of others.

Even in this narrow sense, Autonomy is still a relevant human psychological need. Denial of autonomy is why slavery or incarceration feel so bad. It is also the reason why your two-year-old throws a temper tantrum when you try to interrupt his play in the sandbox.

In contrast to many other media, video games offer a great degree of Autonomy to players. While watching a movie or reading a book, we are passive observers. We might be immersed in the story, but we are not affecting its narrative. Games are interactive by nature. In games, we are the hero of the story. We get to take control over the character and get to make meaningful choices. The player gets to decide which goals to pursue and how to pursue them.

Even the games of chance, where the outcome depends purely on randomness, always employ some means by which the player can have an illusion of control. The act of blowing on the dice before throwing them is an act of asserting Autonomy!

Competence

The second psychological need defined by SDT is centered on what we tend to refer to as skill. It revolves around our desire to control the outcome of our own actions. We have the inherent need to see our actions produce the intended results.

Consider the simple act of throwing a basketball. Before every action comes a moment of intention. Before ordering the muscles to contract and release the ball, our brain forms a mental model of the trajectory he expects the ball to follow. If the result of our action produces the desired result, we feel satisfaction. If, like most times when I throw the ball, the outcome ends up being something random, we feel frustration.

Furthermore, we have a desire to see our games improve based on the observed feedback. The denial of this aspect of Competence is the reason why dead-end jobs can feel so soul-crushing.

The concept of Competence transcends simple physical actions. It applies to our general desire to achieve our goals or to at least get closer to goals with each action we take.

As we have already noted, games are, by definition, interactive. In the context of games, competence is tied to the player's ability to reach goals using the means at his disposal. In this sense, it is closely tied to the concept of Flow popularized by Csíkszentmihályi (Csikszentmihalyi, 1990), which we are going to discuss in one of the following sections of this chapter.

Keep in mind that skills in games can reside on multiple levels. In some games, for example, arcade beat-em-ups, it can be related to dexterity, speed of reflexes, and mastery of controls. In others, such as city builders or turn-based strategies, it is about strategic planning and managing resources. Many games require a mix of several skills. Multiplayer First-Person Shooter (FPS) games require a mix of dexterity and tactical thinking. In other games, skill can be all about social interaction: Game of Poker or the recently popular Among Us hinge on the ability to read minimalistic cues to discern the intentions of other players.

Even in the case of games of chance, players have a tendency to attribute success to their skill, whatever that might be, and failure to bad luck or systemic bias.

Relatedness

Humans are social beings. We evolved through an evolutionary strategy that required group action and interaction. We are meant to be embedded into a social matrix consisting of other human beings. This desire to be connected in some way to other human beings defines our need for Relatedness.

However, this concept is multifaceted. On one hand, we have a deep-seated desire to belong to a group, to feel loved and cared for. On the other hand, we also feel a need to establish our place within the group. To evaluate our skills and social standing in reference to other people with whom we come into contact. The unfulfilled desire for relatedness is why solitary confinement is one of the harshest punishments. Playing board games with your friends, be it as simple as Kimble or as complex as AD&D, is rooted in the desire to satisfy our need for Relatedness.

There are many ways how video games satisfy this need. Since the beginning, video games have offered a chance of playing against friends sitting on a sofa. Atari's Pong, one of the earliest games, was a multiplayer game. Even without a multiplayer mode, arcade machines of the 80s offered a chance to put a three-letter abbreviation on the leaderboard. This rudimentary mechanism was a motivator enough for millions of players.

Many modern games are built around the idea of multiplayer gameplay, FPS and MMO genres in particular. Others have it at the metagame level via various social features, such as clubs, guilds, clans, etc. Even the most casual games have features aimed to satisfy the need for Relatedness. The success of saga map metagame design, made popular by King and its Candy Crush series, is in good part thanks to Facebook friends' avatars shown scattered along the path. This simple hint is enough to help us establish our place in the relative order of our friends. It is a tiny little social matrix for us to feel embedded into.

Even if the game itself doesn't provide means to address relatedness, some of its fans will start to congregate on social media, Reddit pages, and forums to discuss strategies, compare experiences, and simply socialize around the common topic.

Intrinsic and Extrinsic Motivations

Human motivation is a strange thing. I would groan in agony at even a mention of a crunch, yet I would gladly spend a weekend with a couple of all-nighters on a Game Jam. Millions of people around the world scoff at jobs that involve manual labor yet would happily pay each month for the privilege of pumping iron in the gym. From Linux to Wikipedia, much of our modern internet infrastructure is built on software written and maintained by volunteers.

What is the catch? Why do we find some trivial tasks so soul-crushing, yet we persist in grinding at other similar activities?

Understanding the nature of human motivation is a useful tool for any game designer. After all, we are designing games that we expect players will feel motivated to play. This is especially important in the world of modern free-to-play games that are expected to run as a service for years and retain players for months and longer.

In order to unpack the notion of human motivation, we resort to the same psychological framework we discussed in the previous section of this chapter, the SDT (Ryan & Deci, 2000). This theory defines motivation as our willingness to engage in an activity and the drive to persist with this activity until a goal is reached. It talks about motivation in terms of satisfaction of the basic psychological needs of Autonomy, Competence, and Relatedness. In simple terms, people seek out activities that would allow them to satisfy these basic psychological needs. Some activities are better at this than others, and people will gravitate toward them.

Types of Motivation

However, it turns out that there is more than one type of motivation. Broadly speaking, there are two main forms of motivation:

- *Intrinsic motivation*—when we engage with the activity because of the activity itself,
- *Extrinsic motivation*—when we engage with the activity for other reasons.

Intrinsic motivation means that we find the activity enjoyable in and of itself. The SDT claims that this is the case with all activities that are suited to the fulfillment of our basic psychological needs. Basically, we are choosing activities that have the potential to, at least temporarily, increase our mental well-being.

This type of motivation is something of the holy grail of game design. The most successful games are successful precisely because players find them intrinsically enjoyable. In SDT terms, they satisfy one or more, ideally all three, basic psychological needs to a great degree.

A good example of these are games with strong core mechanics. Most traditional arcade and 8-bit games, from Pac-Man and Space Invaders to Super Mario have such properties. Modern examples include multiplayer FPS games and MMOs on one end of the game spectrum and Match-3 or Merge games on the more casual side.

On the other hand, we also choose to partake in the activities, not because of the activities themselves, but because of the broader context. Extrinsic motivation is driven by external factors. Namely by either:

- the promise of a reward,
- avoidance, even fear, of punishment.

This is the proverbial carrot and stick metaphor!

In the gaming world, this type of motivation is usually in the domain of the metagame. All types of reward systems for which players need to endure a period of grind are examples of extrinsic motivation. Consider the mechanic of hatching the eggs in PokémonGO. A player is required to walk a certain distance in order to incubate a virtual egg. In general, a player is not interested in walking 5km without a specific reason. He is extrinsically motivated by the promise of the reward, a new character that will spring out from the hatched egg. Battle Pass systems are typical examples of external motivation.

Even the understanding of the distinction between intrinsic and extrinsic motivation is useful from the point of view of game design. However, the SDT goes even deeper and defines a whole spectrum of different motivational types, as shown in (Fig. 3.2).

The basis of this spectrum is the level to which various activities can satisfy basic psychological needs, primarily the need for autonomy.

Just like darkness is the absence of light, the total absence of motivation is known as amotivation. This is the state of total lack of motivation as a result of a total lack of autonomy! It can be characterized by the feeling of helplessness. A person might feel not capable of acting or compelled to do a particular action regardless of his own opinions or intentions. It can be seen as a lack of perceived value in the action itself or a lack of context.

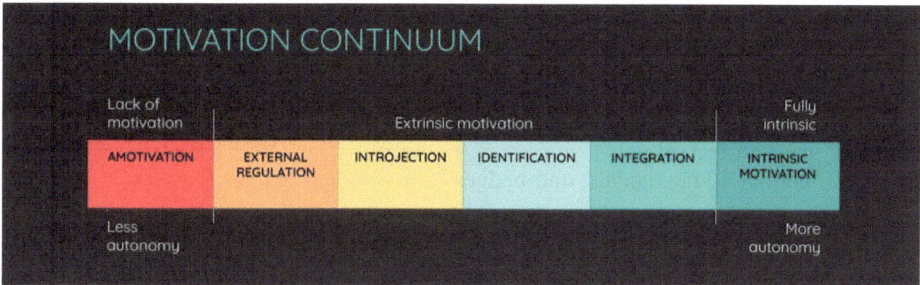

Fig. 3.2 Motivational continuum according to STD

This definition may sound bleak and conjure images of oppression. However, this phenomenon is surprisingly common in games. Examples of UX that can induce the feeling of amotivation, include:

- tutorials in which the player chases the arrow and taps without getting immersed in the content,
- superfluous dialogues in narrative-driven games and cutscenes that players feel the urge to skip,
- "Push button to continue" type of gameplay in many narrative-driven games, where gameplay takes second place to the story,
- grind without context in any free-to-play game.

All of these patterns are something that should be avoided at any cost. These are design failures detrimental to player retention. Put in this situation, players will reassert their autonomy in any way they can, quite often by churning.

The polar opposite of this is, of course, the fully Intrinsic Motivation, which anchors the right side of the spectrum diagram. The area in between these two extremes is occupied by various forms of Extrinsic motivation. These forms differ from each other in terms of how much the action by itself is fulfilling to the player.

The so-called *External Control* is the first of those. It is characterized by the carrot and stick metaphor we discussed already. The action itself is pretty much irrelevant to the player, but the reward is meaningful—the player endures the grind to get to the chest! As such, it is very prominent in modern free-to-play games. So much so that to many players, this is the pattern that defines the whole class of games.

Introjected Motivation is the second type of extrinsic motivation that we are going to examine. It is the part where the ego gets involved and the need for Relatedness starts to appear. The activity may not be rewarding by itself. However, the act of doing it is. This type of motivation is rooted in approval. The approval might be coming from others or even from the person itself.

This is why the bragging rights in games are so important. Especially in competitive multiplayer games. A game designer toolbox is full of design patterns and features that hinge on this type of motivation. To name but a few, this includes:

- leaderboards of any sort,
- achievement systems, medals, and badges in multiplayer games,
- any type of unique items that can be obtained as rewards that can be seen by other players,
- even avatars showing player progression on a saga map in casual games such as Candy Crush.

These types of features can be extremely valuable in boosting player motivation. However, there is a potential pitfall associated with some of them. There can be only one name at the top. Boosting one player's ego can come at the expense of everyone else!

The term *Internalized Motivation* implies that external factors are becoming more and more absorbed by the person. They start to take personal importance in their own right. An individual starts to consciously endorse the value of the action itself.

I do not particularly enjoy the act of sorting out my garbage into categories: glass, plastic, bio, several types of paper, etc. However, I do believe in the importance of recycling.

In the world of gaming, the patterns that correspond to this type of motivation are less tied to concrete types of features and more to the player's behavior. Taking one for the team. Consciously doing any sort of action that doesn't benefit the player directly but matters in the big picture is driven by this type of motivation.

Being a healer in the MOBA team is not directly beneficial to the individual player but is regarded highly by others. Building a replica of the Taj Mahal in Minecraft is another example. Laying down 10,000 blocks is tedious, but being praised for the final result is extremely rewarding.

The final type of motivation that we are going to examine is *Identified Motivation*. This type of motivation blurs the distinction between extrinsic and intrinsic. Participation in activity becomes a part of one's own identity. In other words, the person identifies not only with the value of the activity but also with the activity itself!

In the gaming world, this type of motivation is what makes certain games transcend the mere short-term pastime and transform it into a digital hobby. I am a hard-core Overwatch player. I am a Redstone builder in Minecraft. Some deeply engaged players will self-identify in this way. The presence of this type of motivation usually indicates deep and prolonged engagement with a particular game. The players that are driven by this type of motivation are the game's Superfans! Treat them like royalty!

The Theory of Flow

Do you know that feeling when you are so immersed in what you are doing that you forget about everything else? Minutes and hours fly by. Time loses sense. The reality seems to fade into the background. You are *in the zone*. There is nothing else in your mind, just you, yourself, and whatever it is that you are doing. It could be the feeling you get when playing a familiar tune on the guitar or the joy of riding a bicycle on a warm summer day. It could be that you are writing something and that words are just flowing from your fingertips. It could, damn well be that you are at your work, eliminating your daily tasks one after another. What you are experiencing is the sense of the *flow*.

Even if you have not read anything about psychology and games, you have probably heard the term. It is one of the most widely known psychological concepts that is applicable to game design.

The existence of the state known as the flow was postulated by Mihály Csíkszentmihályi in the early 1970s as part of his research on happiness and creativity (Csikszentmihalyi, 1990). This term describes a particular mental state, one of deep immersion into a particular activity. It is often compared to a related notion of *hyperfocus*. It can be defined as the optimal state of *intrinsic motivation* when the activity itself is so enjoyable and rewarding that the participant does not need any other external motivation to partake in it. We have defined the notion of intrinsic motivation in the previous section of this chapter.

The concept of the flow deals with two additional important concepts: the notions of *challenge* and *skill*. In the context of games, the challenge represents the obstacles that the player needs to overcome in order to attain his goals. The notion of skill denotes the player's ability to overcome these obstacles.

The state of flow is described as just the right amount of challenge compared to the skill of the player. Too much challenge will lead to repeated failures and consequently to the feeling of *frustration*. On the other hand, too little challenge will inevitably result in a state of *boredom*. However, if we are able to present our player with just the right amount of challenge, the player will remain in the blissful state of the flow.

Naturally, when designing a game, we are designing it for a broad audience, i.e., for a wider variety of people. The skill level of individual players will differ greatly from person to person. Furthermore, unless the game is purely luck-based, the more a player palsy the game, the greater his skill will become. Therefore, the skill is a variable that changes both from person to person and over time.

In order to correspond to this, the challenge also needs to be treated as a variable that can be adjusted. These two variables can be plotted on a chart (Fig. 3.3).

The trick in the game design is to adjust the amount of challenge presented to the player so that it corresponds properly to his level of skill. Indeed it is quite common that the difficulty of gameplay grows as the player progresses through the game. This has been the case since the earliest days of game development. Tetris blocks would start to drop faster. Aliens in Space Invaders would start to move with greater speed. Bad guys will

Fig. 3.3 The chart of the flow
with respect to challenge and
the skill

require more shots to be killed, etc. Adjusting the difficulty level of the game in such a
way corresponds to the skill development of an individual player.

Adjusting the challenge level to differences in abilities that occur between different
individuals can be a lot harder. Even in the early 16-bit era, games started offering an
option to choose the difficulty level of the game at the very beginning. Figure 3.4 is a
screenshot of the difficulty selection UI from the original Doom by id Software. This is
still a common design trope in many single-player, skilled-based games.

Controlling the ratio of skill to challenge is relatively manageable in single-player
premium games, in which the game designer controls the gameplay experience throughout

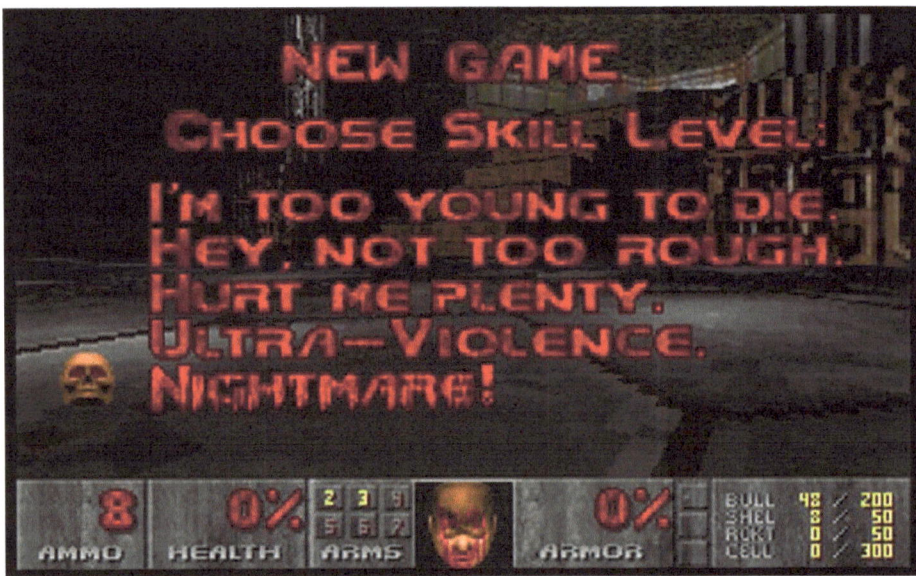

Fig. 3.4 Skill level selection menu of the original Doom, by id Software, 1993

the player's journey from the very beginning until the end. Things become a couple of orders of magnitude more complex when it comes to free-to-play games or multiplayer games.

Free-to-play games are designed to be open-ended and are meant to retain players for months or even years. Neither the player's skill nor the difficulty of the game can be expected to grow to infinity. Keeping them parallel, i.e., keeping the two skill curves and the challenge curve on the flow graph from crossing or diverging, is one of the greatest challenges when designing games for long-term retention.

Multiplayer games, regardless of whether they are premium or free-to-play, rely on the participation of multiple individuals. These individuals will, of course, have different skill levels. Noobs and pros do not mix well in such environments. Over the years, a whole slew of matchmaking methods have been developed to address this problem, especially for 1v1 games. A good example of this is the Elo rating system developed for matchmaking of players in the traditional game of chess (Elo, 1967). We are going to discuss the Elo rating system in more detail in Chap. 6 of this book.

Things get even more difficult when it comes to games in which teams of players compete against other teams of players. Matchmaking multiple individuals with different skill levels and keeping the challenge appropriate is by no means an easy task. Microsoft's Trueskill system (Herbrich et al., 2007) (Minka et al., 2018) is an extension of the Elo rating system designed specifically for this purpose.

Designing for Emotional Engagement

Imagine a scene… *Golden rays of sunlight are shining through the window. Tiny particles of dust are glittering as they float slowly through the air. There is a faint sound of children giggling outside, somewhere in the distance. The air is warm, and so is the soft quilt covering the bed. A faint smell of old paper and printing ink is filling your nostrils. The glass of soft drink feels presently cold as you grasp it in your hand. The ice cubes jingle gently. The soda bubbles pop and hiss…*

Imagine another scene… *The neon light keeps on flickering. Its cold cyan glow is irritating. It is too faint to properly light up the room in which you are, yet it is intense enough to hurt your eyes. You can see a wall. Cold, hard, unyielding concrete, painted pale grey with a sickly green hue. The floor is covered with white ceramic tiles. Dry stains of coffee-colored splatter lead to a drain, a dark, round, gaping hole in the ground covered with a metal grill…*

Small hints can evoke powerful emotions. The sound of a can of Coke being cracked open… The hue of light in the late summer afternoon… A visual and sound cue can set the mood of the scene. The words alone can create a sense of immersion into the setting. It's a cheap trick, but it works!

These small elements are important building blocks of narration, helping to create an emotional bond with a reader of the book, a viewer of the movie (Xiangyi, 2016), or, indeed, a player of the game.

It is this emotional connection that you, as a game designer, should seek to create. People play games to be entertained. We feel a sense of fun on a deeply emotional level. Emotions are a way by which our subconscious mind tags experiences, marking them as important or fulfilling, i.e., something worth remembering.

The connections created on the emotional level are much stronger and more long-lasting than any of the memories of particular details. You might have forgotten the details of that Nintendo game you loved so much as a kid, but you still get that warm and fuzzy feeling when you think about it.

However, many game designers, especially when it comes to free-to-play mobile games, seem to forget this simple thing. If your strength is in the game systems design, and you focus on the retention of players, you can easily drift into creating mechanistic designs, a set of rules that might work well on paper or in Excel spreadsheets but is utterly devoid of soul. I admit this is a pitfall that I myself have fallen into too many times.

Recently, I stumbled upon a 2017 whitepaper on the design of so-called cozy games (Short et al., 2017), and it was an eye-opening experience!

While I am fascinated by the concept of cozy games itself, it is the systematic approach that the authors took to define the concept that really caught my attention.

The authors begin by defining a feeling that they want to focus on. They define a specific emotional state that they want their game to elicit in the player. In their case, this is a sense of coziness, described with a set of adjectives such as safety, softness, and abundance.

Afterward, they specify a list of emotions that partially overlap with the concept of coziness, for example, such as cuteness, memories of childhood, romance, etc.

In the next step, they proceed to list a set of notions that stand in contrast to the feeling they want to evoke, emotions like fear, sense of danger or threat, design patterns such as extrinsic rewards or artificial resource scarcity, etc.

Finally, they proceed to discuss the means by which a sense of coziness can be evoked in a player of the game. They list the relevant:

- General aesthetic patterns of coziness,
- Game design patterns and mechanics that reinforce the feeling of coziness,
- Cozy visuals and audio elements,
- Cozy narrative patterns,

They even proceeded to list the items, locations, and character archetypes that can be characterized as cozy.

In this way, they are actually creating a sort of game design moodboard. By doing this, they put the desired emotion in the center of their design process!

This moodboard is a reference to which they can always return to evaluate their own game design ideas. Furthermore, they define a set of tools and a palette of game design elements that all serve to reinforce the particular feeling that their game is supposed to be centered on. The whole game, it's every aspect, is designed to reinforce the same general mood!

In this particular whitepaper, the authors focus exclusively on a sense of coziness. However, the same approach can be used for any type of emotion or complex sensation.

People play games to be entertained, but they can be entertained in multiple ways. Just like a book or a movie, a well-crafted game can elicit all sorts of emotions in the player.

Building Your Own Moodboard

You can try to replicate a similar process and apply it to your own game design needs.

Begin by selecting a particular mood or feeling that you want your game to focus on. For example, creepiness. Try to put the definition of this sensation into words. For example, creepiness is eeriness, a sense of unseen danger, a sense of dread, a feeling of unease, etc.

Discuss with as many people as possible to dig for ideas and reach a common understanding of the emotion you are trying to elicit. Different people will find different things to be creepy, but there is bound to be a huge overlap. Clowns, for example? Right?

Define contrasting feelings. The antithesis to creepiness can be a sense of security, pleasantness, coziness, etc. List overlapping and adjacent concepts, the ones that share some traits with your target but are yet distinctly different. In the case of creepiness, these can include gore, terror, existential horror, nausea, a feeling of helplessness, the feeling of being lost, etc.

Proceed to define the aesthetic elements:

- Creepy visuals: darkness, dim, faint light, cold colors, etc., (Fig. 3.5)
- Creepy materials: stone, metal, ceramic tiles, rust, wet surfaces, cold surfaces, reptilian skin, viscera, etc.
- Creepy audio: the sound of creaky floorboards, the sound of footsteps in the distance, the sound of heavy breathing, the sound of one's own heartbeat, the sounds of insects, the sounds of hissing, the sound of rumbling machinery, etc.
- Creepy locations: abandoned buildings, empty streets, crypts, cemeteries, abandoned amusement parks, old hospitals, derelict bathrooms, narrow corridors, confined spaces, old castles, ruined churches, dark forests, etc.
- Creepy items: broken toys, broken baby strollers, old photos, chains, medical instruments, hospital beds, etc.

Fig. 3.5 Photo by Yener Ozturk

Use this list of elements as a reference to evaluate your design whenever a new design element is added, or use it as a source of inspiration. Let the artists in your team use it to create an actual art direction moodboard to be used to define the visuals.

Keep iterating and improving this list as you go by adding to it to refine and reinforce the general direction in which you are heading.

This approach is explicitly designed to be holistic, where every aspect of the game is there to reinforce the overall experience. In this way, you can circumvent the common problem of soulless games.

In addition, because all aesthetic aspects of the game work in unison, you do not need to rely on conveying the narrative of your game in terms of storylines and dialogs. The narrative will emerge from the setting of the game and can be reinforced by minimal means of snippets of dialogue or text.

Ideally, taking this approach would result in a more coherent game in terms of the emotion that it aims to create. Emotion should result in a deeper connection with the player.

Of course, this is not meant to be the ultimate design method. It is just a tool that you can use in conjunction with all other tools in your game designer toolbox. Combine this with some solid game system design, and you should be on to a winner.

Perceiving Randomness

In Finland, where I live, gambling is legal, and slot machines are in every shopping mall. They are usually played by senior citizens, but I knew a young guy who had a foolproof plan of how to win. He would lurk around the machines, waiting to spot another player giving up after a long sequence of futile attempts. The guy would then jump on the machine that was free now and start to play, sure that after a long sequence of lost games, the machine was bound to throw a winner. Eh, he never won big, and he lost more than he had won, but he still remembers only those times that he won anything. I bet he is still waiting for his big win, which surely, after all these years of playing, must be coming any day soon. Right?

Of course, the math doesn't work that way. The probability of you hitting a jackpot doesn't actually grow each time you lose. Still, good luck in trying to explain this to some people. The reason why people believe in this stuff is profoundly tied to the way our brains perceive randomness. We perceive randomness in a way that is fundamentally flawed yet rooted in a solid survival strategy.

Video games can harbor randomness in various places. As a game designer, you will eventually be going to use it, so it pays off to understand how players are going to perceive it.

Randomness in Nature

The brain is a pattern-recognition organ (Mattson, 2014). Any brain in any kind of animal. The purpose of the central nervous system is to ensure the survival of the organism. We, like any other living creature, live in a chaotic, random world. This randomness is driven by something fundamental to the nature of the universe, and we do not quite understand it (Aaronson, 2014). In order to deal with the ever-changing environment, living organisms have adopted two broad strategies. Some, like plants or fungi, are happy to stay anchored to one spot, ready to suffer whatever calamity the elements throw at them. Others have chosen to stay mobile, constantly searching for a source of food or trying to get away from potential danger.

These movements are inevitably guided by some sort of nervous system. Boiled down to its most basic essence, there is only one question that even the brain of the simplest animal on this planet needs to answer at each moment in life: Will things be better if I move over there than if I stay over here? In order to answer this question, a brain needs to be able to predict the future. Not in the sense of guessing the movement of the stock market but in a sense in which a sparrow brain is able to predict where the tier of an oncoming car is going to be in the next three seconds.

In order to do so, the brain needs to spot patterns, patterns of movement of objects, patterns of colors, patterns of lights and sounds, and patterns in any sensory input. Brains hate randomness. True randomness has no pattern. There is nothing to be learned from randomness. Quite the opposite. For a creature with a brain, it is better to assume that there is a pattern where no pattern exists than to assume that things are just chaotic and miss picking on an actual pattern of behavior. That whisper in the grass could be just the wind. It could also be a cat lurking.

The human brain is exactly like any other brain on this planet, only more complex. This is the reason why phenomena such as pareidolia exist. This is why we see faces in electric sockets or Jesus on a piece of toast. There is an evolutionary advantage in assuming that a pattern exists even in really random input. Humans are hardwired to perceive patterns even if no pattern exists!

Randomness in Games

Obviously, this has some very fundamental implications on how we humans perceive randomness, especially randomness in games. Imagine that you are rolling a die, a simple six-sided die. The chance of you rolling 6 is always, and at each roll, exactly 1/6. This is what is known as an independent and identically distributed random variable (Glen, n.d.).

However, an average human will assume that the chance of you rolling 6 increases each time that you roll any other number. This is known as Gambler's Fallacy (Oppenheimer & Monin, 2009). Our brain assumes that there is a pattern. It is hardwired to do so. It will assume that the chance of rolling a particular number in a particular roll is somehow tied to the outcome of several preceding rolls. This is why some people believe that there is some magic system by which you can fill in Lotto numbers or increase your chances at the lottery.

Dice, bingo, and lotto are simple, they are straightforward games of chance. Videogames, on the other hand, can hide randomness in many places. If the pattern of the outcome of a roll does not correspond to the player's expectation, the player will assume that the system is deliberately trying to sabotage his efforts or withhold something from him.

Assume that you are making a game in which a player, from time to time, gets to open a chest. There are three possible items that he can find in each chest: a healing potion giving him an extra life, a piece of gold, or a bomb that can be thrown at enemies. If not otherwise communicated, players will assume that the probabilities of finding any of these are equal, i.e., 1/3. They will also assume that after opening, for example, 30 chests, they will have about 10 of each item in their inventory. If you design your system to be truly random, what is likely to happen is that an average player will end up with a big surplus or a big deficit of one of the items. For example, an inventory might hold 4 healing potions, 6 pieces of gold, and 20 bombs!

The value of these items in a game can be very contextual. For example, gold is cool when you are standing in front of a merchant selling weapon upgrades, but to reach that spot you need to clear a forest path of a pack of trolls, and those bombs would come in handy in that situation. Players' brains will inevitably imagine a pattern in the randomness of this system. A player running away from a horde of baddies with only half a heart, desperately searching for a healing potion, will inevitably think that the system is deliberately trying to screw him over after opening the fifth chest only to find another lousy piece of gold. A while later, the same player might be searching for gold cos he has 997 pieces, and that slick new sword costs 1000 at that merchant's place, and all he can find is just healing potions and bombs! Surely, the system is again deliberately hiding from him just the things he needs! Humans expect conditional probabilities even if outcomes are independent.

There are several strategies that you can use to mitigate this problem. One obvious strategy is to implement a system that would take into account the player's expectations. For example, you could increase the likelihood of a player finding a healing potion when his energy is low or the probability of finding bombs in chests that are close to spawning spots of enemies. This is a traditional method employed by many games for years. The pitfall of this method is that you might allow for an easy exploit. If players figure out the set of conditions that makes a certain outcome more probable, they might also be able to replicate them at will, thus breaking the balance of your game.

The other approach is to give an illusion of control over the outcome into the hands of the player. No one is blaming dice for not rolling 6 simply because they are the ones doing the actual roll. Many games include this option. The player is required to select a box or a card or to choose which crate to smash. This is pure smoke and mirrors, and the success of this psychological trick depends very much on the minute details of UX and visual presentation.

You can also try to balance out the amounts of items that the player will eventually get. For example, you could generate a list of 30 elements with 10 of each three item types and randomize the order in which they appear. You can use this list to decide the outcome of opening a chest. This will ensure that after opening 30 chests, the player's inventory will contain an equal number of health potions, bombs, and pieces of gold, resulting in a more fair game. There is no universal solution, and the choice of approach depends on what kind of game you are trying to make.

Randomness of Action

Remember the Self-Determination Theory. It is a psychological framework dealing with human psychological needs. One of three fundamental needs is characterized as competence. Simply, we as humans want to see that when we take some action, the outcome of that action will produce an expected result. For example, when I kick the ball, I hope it will go in the intended direction. If the ball instead veers off in some random direction, I will feel frustration.

This applies to the world of video games as well. If the outcome of the player's conscious actions is something random, this is likely to cause frustration. If randomness is tied too closely to the player's actions, it will cause frustration. If randomness or the outcome is not hinted at in some way it will cause frustration. For example, a player will feel fine if he opens a chest with a big question mark painted on it and a random thing happens. He will not feel fine if he throws a hand grenade and it flies off in an unexpected direction. People want to fight randomness and overcome it using their skills!

What players want to see instead is a random change in the environment to which they can react using their skills. The outcome of their actions needs to be predictable and in accordance with their expectations. Otherwise, they will doubt their skill and the fairness of the game and start to feel frustrated.

Perceiving Patterns

The purpose of the brain is to predict the future. This statement of mine might terrify you with its boldness, but bear with me for a moment. This is not your garden variety tin-foil-hat theory, although my story has much to do with them. Also, I use the word "purpose" here for the rhetorical effect only, as there is probably no particular "purpose" to the existence of any particular organ.

When I say predicting the future, I don't mean predicting who will win the lottery next month or what will be the score of next year's Super Bowl or anything so silly and random. The function, one of many, of the central nervous system in any creature equipped with it is to ensure the survival of its owner by predicting the future. For a frog, that future might be two seconds in advance, and survival might mean determining the size, speed, and direction of the movement of a large incoming object, ensuring the timely jump out of harm's way, Or it might be telling apart potential prey from inedible objects. A mental calculation that goes something like this. Oh, there is something small flying this way. It looks like a fly. It might be tasty. If I stick my tongue out in 3.25 s, it will intercept its path, and I will catch it. The frog does not know anything about seconds, degree angles, miles per hour, or any such thing. Not consciously, at least. But a frog's brain does. It can do this calculation in a fraction of a second. Over a couple of million years, it has evolved to do precisely this sort of calculation. It's a survival strategy. Of course, you don't need a brain to survive, and I don't have reality TV stars in mind when I say this. Plants, mushrooms, and sponges live their entire lives happy and brainless. Still, it's a survival strategy chosen by a sort of lifeforms to which we, ourselves, belong.

We, humans, are a strange bunch. We've pushed this strategy to the extreme. Our brains are the best weapons in our fight for survival. The motion of a swamp fly, the shape formed on a frog's retina by light reflected from the fly's body, is a complex pattern. Predicting its change in the next ten seconds is not a small feat. However, our human brains deal with patterns that are several orders of magnitude more complex. For us, the time window of prediction might be not mere seconds but days, months, and years up in advance. This is something that defines us as sapient creatures, distinct from other animals, that sets us apart even from our closest primate cousins.

There is a river in France. Archaeologists have uncovered the remains of two distinct Paleolithic stamens on its banks. One belonged to Neanderthals and one to Homo Sapiens Sapiens, the Cro-magnons, our direct Stone Age ancestors. They can tell them apart by the subtle difference in tools, in the way the flint blades were chiseled out of the rock, but the main difference between them was that Cro-magnons one was used only occasionally, and the Neanderthal was permanent. The two camps coexisted side by side for centuries. The river was full of salmon, a prime source of nutrients. That is why they were there. Yet salmon swam up the river only once a year, a regular yearly migration circle. And that's the catch. They all knew that salmon would eventually come, that there is going to be fish in the river, but Cro-magnons knew their time patterns better. They knew when the fish would come back. They knew they had the time to go and pick berries and hunt rabbits in the meantime and come back to catch them at the right moment. Neanderthals had a different sense of time or none at all (Petru, 2017). They had to sit there and wait for the fish lest they would miss the moment when the salmon returned, missing all the hunting opportunities away from the river.

Being a smart kid in high school doesn't mean being able to answer all the questions on an ABC test. It means being able to realize that if I don't mess around now and sit and learn this stuff, I will not get in trouble three months later when the test comes. There is a so-called marshmallow test, a long-term psychological study that shows that people who, at the age of three, were able to longer resist the urge to eat a marshmallow that was given to them did better in life later (Mischel & Ebbesen, 1970). The size of the prediction time window is, therefore, a measure of intelligence.

These patterns are one or more variables that change their value over time. Mathematics calls them time series. The prediction process starts by gathering the data and observing the variables and how they change over time; then, one can hope to guess how the same variables will behave in the future and which values they will take.

Since this process is essential for our survival, we are hard-wired to do it. We are hard-wired to even joy doing it. Our brains squirt out a bit of dopamine whenever we manage to make a good prediction, a small reward for doing a good job. More precisely defined, the brain is a prediction error minimization organ (Millidge et al., n.d.). It constantly makes predictions about time-based patterns, observes the results of its own predictions, and rewards itself with a hit of dopamine if things go as expected.

Music is just one such pattern, complex yet regular in some way, predictable. This is why we derive pleasure from listening to it. We learn to recognize patterns in the same way machines do by examining the patterns that we know over and over again. This is why little kids like to watch the same scene in their favorite cartoon three hundred times in a row and why some people like to watch cliche stories in movies. But the mechanism works in both directions. Our brains know when there is nothing new to learn, i.e., when reexamining a pattern is not effective. This is why some people find predictable stories boring. The threshold is a personal thing. Some people learn quickly and find things that are easy to predict more boring. Maybe there is a correlation between IQ and the complexity of music and the storyline we find entertaining.

Prediction is easy if the interplay between variables that describe the system is simple. These types of systems are known as linear. Our brains are really, really good at doing this sort of prediction. This is pretty much the only type of prediction we can do. We are wired to seek out such patterns, and we see them even if they don't exist. Nature errs on the side of safety. Thus, pareidolia and the face of Jesus on a piece of toast (Wardle et al., 2020).

But what if a system is more than a sum of its parts if its behavior is not a simple sum of the values of its variables at any given moment in time? A small change in one variable can cause a disproportional change in the values of other variables in the system down the line. The proverbial butterfly effect of chaos theory. The weather, the stock market, and our own metabolism are all such nonlinear systems. This is why we are not able to predict Wall Street or the result of next year's Super Bowl.

The catch is that even such systems behave in a linear way in some time range or range of variable values. We can predict the weather three days in advance. This is why our whole evolutionary mechanism is useful even in the nonlinear environment that we inhabit. We can't do these nonlinear predictions. We are very bad at them. Drinking beer from a curved glass makes you get drunk more quickly (Attwood et al., 2012). Our brain mucks the prediction of the time it will take you to drink a pint. This is not just an imperfection of our brains. Mathematics breaks at this point (Schittkowski, 2002). There is no easy way to handle it, and we are terrified by it. We would like the world to be predictable. Small changes that result in gigantic events scare the hell out of us. This is why conspiracy theories seem so appealing. Big events need to have big causes, even if we do not see one, surely it must be a conspiracy (Douglas et al., 2019).

Target Audience

4

Understanding your audience, i.e., the people you are making the game for, is one of the essential prerequisites of a good game design. This is actually true for any design process, be it as mundane as a new kind of pencil or as complex as designing a telesurgery system. The approach that puts the user in the center of the game designer's attention is known as the *user-centric design*. Its equivalent in the world of gaming is known as the *player-centric* game design.

We have begun our journey in this direction already in the previous chapter of this book. You will notice that both the previous and this chapter deal a lot with the psychology and the motivations of the players. However, there is a good reason for having these related topics divided into two separate chapters. In the previous chapters, we discussed the underlying psychology that applies universally to all humans, i.e., across the whole potential player base of a game. In this chapter, on the other hand, we are going to take a look into parameters defining distinct subsections of players. In other words, we are going to discuss ways of segmenting the player audience.

This is important for one very simple reason. Despite sharing the basic psychological needs and motivational framework, human psychology can be incredibly diverse. The player audience of any game is never going to be a homogenous mass. Inevitably, some groups of players will have radically different attitudes and behavior from others. Understanding your audience means acknowledging the existence of these groups and understanding their motivations and particular needs.

There are two general approaches to modeling the target audience: the *top-down* and the *bottom-up* approach. The choice of approach is determined by the phase in which the particular game development project is at the moment. Modeling of the target audience is often done at the start of the game development project, during the preproduction phase,

S. Stanković, *Game Design for Free-to-Play Live Service*, Synthesis Lectures on Image, Video, and Multimedia Processing, https://doi.org/10.1007/978-3-031-56156-6_4

when the potential game idea and the concepts are being evaluated, and the potential product market fit is still examined. In this case, the *top-down* approach is taken. This means that a lot of assumptions are made about the potential ideal audience and the player profiles that the game would be aimed at.

On the other hand, in the live service phase of the project, when the design process is focused on providing new content and features for an already existing game with an established player base, a *bottom-up* approach is taken. In other words, player profiles are built using the accumulated data that is available about the player's behavior within the game.

Demographic Data

One of the traditional ways of modeling the player audience is using demographic data. This data can be derived in various ways. For example, age verification UIs provide somewhat reliable information about a player's self-reported age. Other tools for the collection of demographic data can include optional surveys or data shared by players using connected social media accounts. Typically, player age, gender, and geographic origin are used for this purpose.

This is a good point to share some general demographic data about the population that plays video games in general. According to the 2022 survey collected by the Entertainment Software Association (ESA) (Electronic Software Association 2022), 71% of children and 67% of adults, i.e., persons at least 18 years old, in the US play video games. Adults make up over 76% of all players. The average age of a player is 33. The age distribution is as follows: children under 18 make up 20%, people between 18 and 34 years old are the single biggest age group with 36% of all players, age groups of 35 to 44 and 45 to 55 make up an additional 13% and 12% respectively, players older than 55 make up the remaining 15%, see Fig. 4.1.

Nearly one-half of players, or about 48% of the players, identify as female and 52% as male. Over 70% of men and boys and 62% of women and girls play video games for at least an hour each week. The situation is not significantly different globally.

As we can see, video game players, in general, cover a relatively large demographic block. However, things might be very different when it comes to particular games. For example, the gender split of PlayerUnknown's Battleground (PubG), according to estimates (Clement 2022), is going to be much more male-dominated, with 57.4% male players versus only 42.6% female players. The picture is even less balanced for a game such as Design Home Makeover, a game explicitly marketed to women, which has a 75% female versus only a 24% male audience (Knezovic 2023); see Fig. 4.2.

The player distribution per age group can also vary significantly from game to game. Fortnite by Riot Games is a good example of this (Ruby 2023), with over 62.7% of players younger than 25.

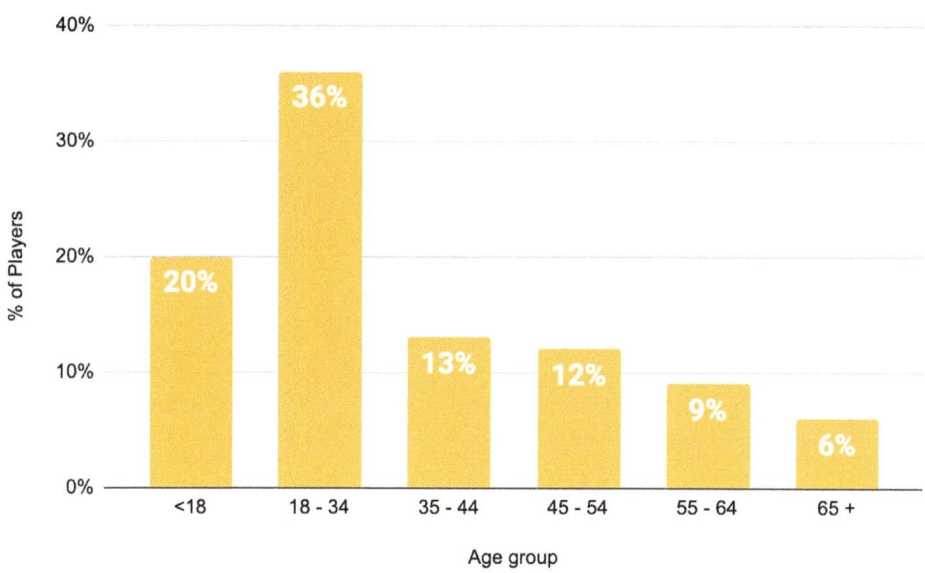

Fig. 4.1 Distribution of players by age groups in the US

It is easy to overestimate the importance of the demographic data. Back in 2009, when it launched, the original Angry Birds was the first major breakthrough hit on their brand-new smartphone platform. Even the mainstream press was writing about it. Millions of people purchased their shiny new iPhones. They wanted to play some games and went to install the only game everyone was talking about. Rovio naturally wanted to learn about its audience. The data showed that their games were played by pretty much everyone, as they were the best-known games in a small market. Diligently, Rovio concluded that this was an inherent quality of their own games and not a quirk of the marketplace. Accordingly, they concluded that this was the way to repeat the success and spent the next couple of years chasing the so-called *Four Quarters* games (4Q), meaning the games that would appeal to all four quadrants of the chart, the young, the old, the male, and the female. In the meantime, the competition emerged, filling the market with games that appealed to specific subsets of the players. Rovio's audience inevitably evolved and became increasingly younger and more male. At some point around 2013, Rovio decided to produce games targeting specific demographics. Angry Birds Stella was launched in February 2014, targeting specifically the teenage girl audience (Clarysse 2014). Despite the best efforts by Rovio, see Fig. 4.3, the typical AB Stella player was still a 9-year-old boy, and the gender split remained 60% male versus 40% female (SensorTower n.d.).

The age distribution data can be more helpful. Typically, the age of the players reflects their available disposable income, their daily habits, etc. Arguably, the younger audience has less money to spend but more time. The competition for this segment of the market is

Fig. 4.2 The gender
distribution of PubG versus
Home Decor players

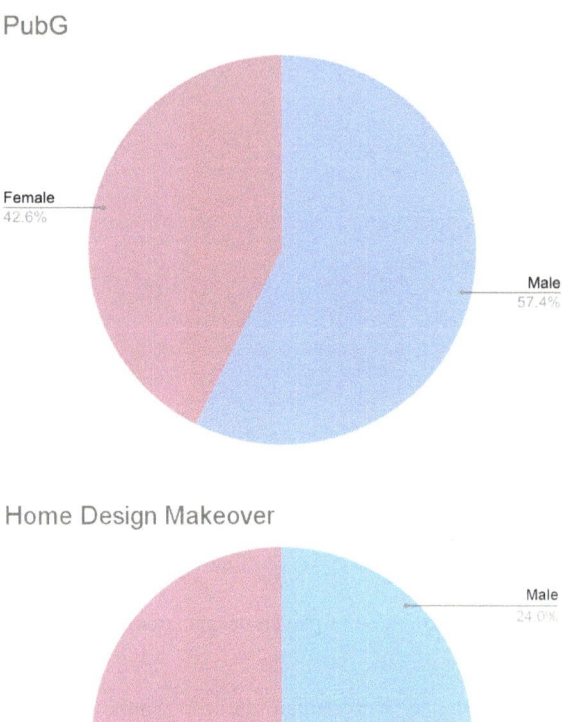

also stiffer as more games are targeting this audience. According to data (Statista 2023), the age group of 34–45 years old has the highest household annual mean disposable income of around $97,000.

Demographic data about age and gender is often augmented by geographic data. Globally, the majority of players reside in the Asia–Pacific region, or 54%, followed by Africa and the Middle East with 16%, Europe, Latin America, and North America with 13, 10, and 7% of the global player population, Fig. 4.4, (Wijman 2023).

The knowledge about the geographic distribution of your players can be very valuable. From a purely business perspective, it can help shape the monetization strategy of the game and determine the pricing policy of IAPs within the game, as players in different regions can have radically different spending patterns. On the design level, it can help select and create culturally relevant content for the game. This can be especially important when it comes to planning the live service. For example, if your game happens to be

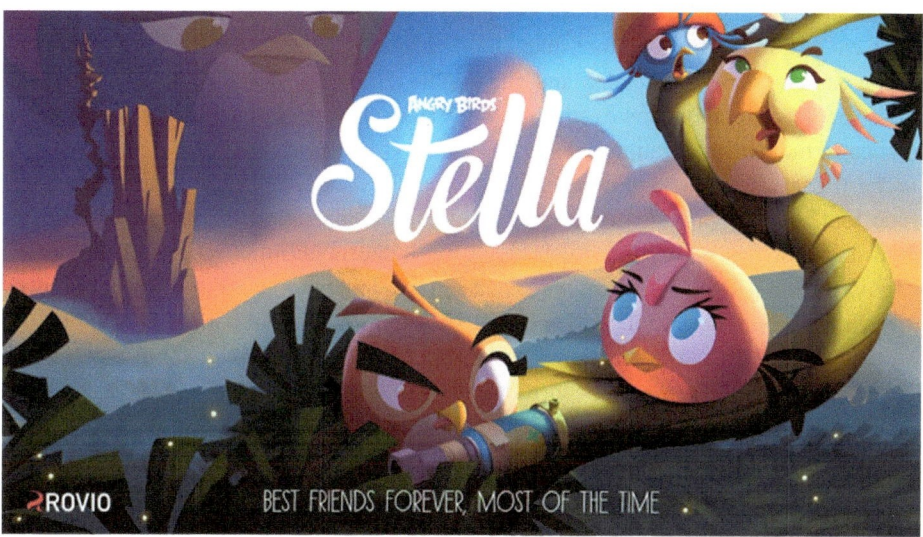

Fig. 4.3 Artwork for angry birds stella by Rovio

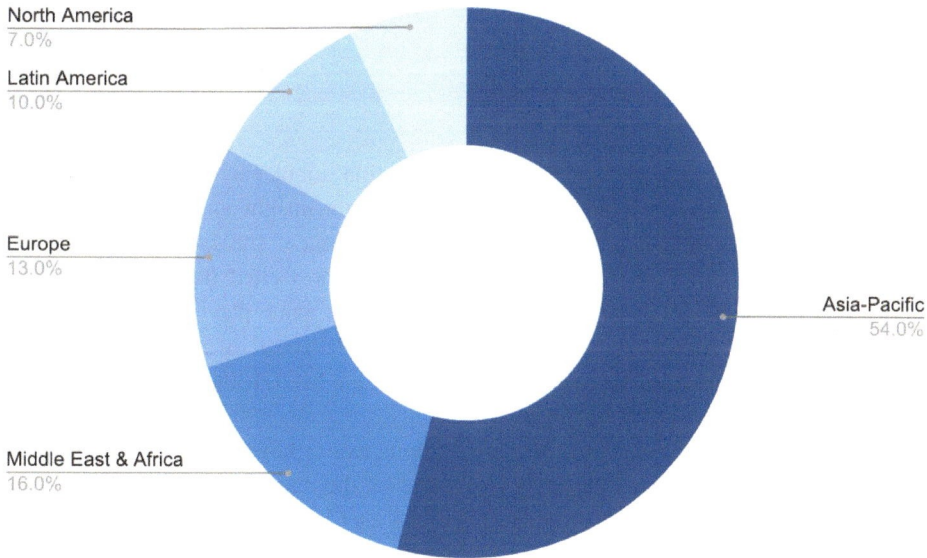

Fig. 4.4 Global distribution of video game players

popular in a country like Mexico for example, it might make sense to invest in creating a live event tied to the local culture, like Día de Muertos.

You can augment age, gender, and geographic data with additional demographic categories, such as average income, education level, etc. However, this is not a very common practice in the world of gaming.

The biggest disadvantage of using demographic data is that it remains a somewhat crude tool. It does provide some insight into the composition of the player audience, but it does not necessarily tell much about the interests and motivations of the players. A famous example of this is the meme that made circles around the Internet about two elderly males, both born in the UK in 1948, both having two children, both being married twice, engaging a lot in international travel, interested in cars, and owning a castle. One of them being King Charles III and the other one being Ozzy Osbourne (Ward 2016). I wonder, would those two be interested in the same game? Clash Royale, perhaps?

To summarise, in the gaming industry, the gender, age, and geographic origin of players are the most demographic data. The gender of the target audience can provide insight into the preferred genre and the art style of the game. The age distribution of players can provide information about potential available disposable income, spending patterns, and genre preferences. Finally, geographic information can help shape monetization strategy and pricing policy as well as help create culturally relevant content.

Bartle's Taxonomy

Demographic data does not provide direct information about player motivations or the gameplay patterns for a specific game. In order to gain insight into these matters, we need to apply other methods of modeling the player audience. The so-called Bartle's taxonomy of players is one of the earliest and, thus far, most cited frameworks designed specifically for the purpose of the classification of video game players. It was first proposed by Richard Bartle in his 1996 essay (Bartle 2004). Bartle is a game designer and a multiplayer game pioneer. He was the co-author of the MUD1, the original Multi-User Dungeon.

He used his experience to divide the players along two orthogonal axes labeled Acting-Interacting and Player-World, respectively. Plotted on the diagram, these two axes create four quadrants, see Fig. 4.5. Players belonging to each of these quadrants are denoted with the four labels as *achievers*, *explorers*, *socializers*, and *killers*. As a mnemonic, each of these player archetypes was associated with a card suite of diamonds (♦), spades (♠), hearts (♥), and clubs (♣), respectively.

Each of these player archetypes is characterized by their attitude towards other players, the game world, and the affinity of either action towards other entities or interaction with them.

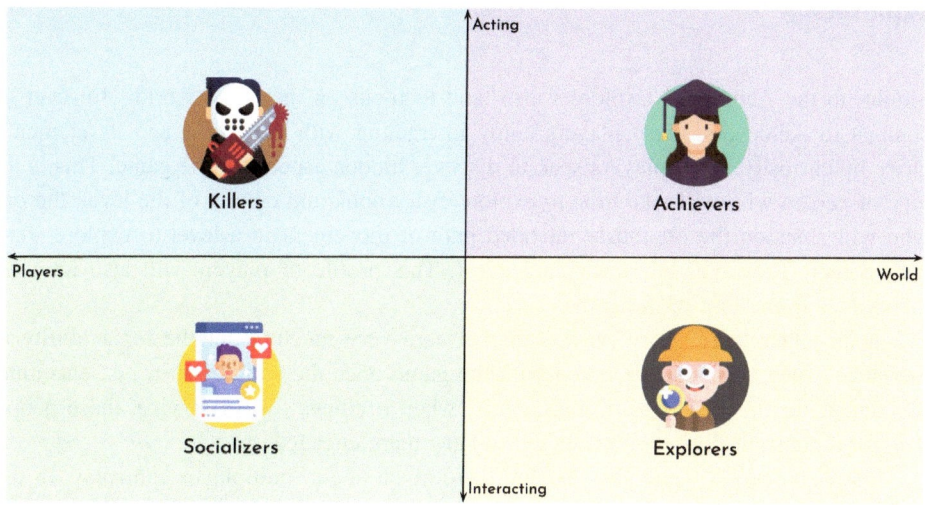

Fig. 4.5 A chart of Bartle's taxonomy of players

Achievers ♦

These are the players who prefer to focus on the gameplay world and act against its elements. They are driven by personal success and achievement. These are the people who like to tick the checkboxes. Typically in the context of single-player games, these players are going to focus on reaching the ultimate goal of the game, its end, i.e., beating the game. They are also going to be focused on collecting any kind of points, tokens, or other external indicators of success. They enjoy completing various tasks and quests. In addition, they would enjoy completing any sort of collection. These are the people who would share videos of game speedruns.

When it comes to multiplayer games, these players strive to show off their skills. They enjoy collecting various trophies, badges, and similar visual displays of status and achievement. They depend on other players, socializers especially, to derive social approval. Typically, they will strive to unlock and max out every character within the game.

Achievers typically get frustrated by open-ended games or any game without clear progress indicators. They require clear goals and external signals of success. They enjoy the replayability of games as long as they can strive to reach new achievements.

Explorers ♠

Similar to the Achievers, Explorers also tend to focus on the game world. However, in contrast to Achievers, these players enjoy interacting with the world and its elements. Drive by curiosity, these players seek to discover hidden aspects of the game. This is the type of person who will take time to explore every nook and cranny of the level, the one who will abandon the obviously intended path of movement in a level to explore some dark corner. They will take every sidequest. This profile of players will also immerse themselves in the lore of the game.

On the other hand, these players do not care very much about the replayability of the game. They typically get bored with the game once they consume it, i.e., encounter its content for the first time. Consequently, when it comes to live service, these people demand a constant flow of new content to keep them engaged.

This player profile generally does not care much about multiplayer gameplay. In this context, they will seek to explore, i.e., unlock all the possible characters, skills, arenas, and locations. They will especially seek different gameplay experiences within the same game.

Socializer ♥

The heart symbol is an apt representation of this player profile. These are the people that thrive in the company of others. They seek out multiplayer gameplay experiences specifically to interact with the other players. They are often the glue that keeps the game's community together. These are the players that offer social approval and admiration to achievers and the satisfaction of dominance to killers. In the team setting, they will selflessly take on the supporting roles. This player archetype typically provides the rank and file of guilds within the games. If offered the opportunity to rise to the ranks of guild leaders, they often tend to be facilitators and arbiters of collaboration and enforcers of fair play rules.

They are generally not attracted to single-player games, but if they do, they tend to gravitate towards hot and popular new titles that offer opportunities for social interaction outside of the game itself, but rather a conversation about the popular game with other people.

They are the least result or outcome-driven of all four player archetypes. For them, the experience of socializing is the outcome.

On the other hand, they tend to get frustrated if unwritten social norms are violated. In a runaway scenario, the guilds run by such players can turn into nightmarish lord-of-the-flies dystopias.

Killers ♣

Killers are a stark contrast to socializers. They, too, require the company of others, but unlike socializers, these players strive to show dominance over others. Consequently, they, too, gravitate towards multiplayer games, but only as an arena to show off their skills. Similar to achievers, they will strive to climb the leaderboards and collect the trophies, but unlike achievers who want to attain rank to show how good they are in general, killers strive to show how much better they are in reference to other people.

The so-called Bartle Test of Gamer Psychology is a questionnaire accompanied by a scoring formula designed to determine the dominant player personality of a person (Bartle 2004). There are several implementations of this test that can be found online.

In general, they enjoy a fair fight and a worthy opponent, but many of them will succumb to the dark side and enjoy humiliating other players whom they deem inferior. They get frustrated by losing, especially by losing to other players. They are prone to ragequitting. They are also a player profile most likely to resort to cheating. If they cannot win by legitimate means, they will seek a way to bend the rules to gain an edge over the others.

Bartle's taxonomy is a useful tool in modeling player profiles. Above all, unlike other psychological models, it was tailor-made specifically for applications in gaming. However, it too has many shortcomings. It does try to fit the whole multidimensional spectrum of human psychology into rigidly defined categories. Furthermore, these categories are defined based on just two vectors. This is often quite limiting. Other psychological models often offer more granularity in modeling a player's psychology and offer more insight into player motivations and expected play patterns. Probably the best way of using this particular model is by treating these four categories as variables. In a way, each player, i.e., each person, has inside of himself a bit of all of these four player profiles, with one perhaps being dominant.

Big Five and OCEAN Model of Personality

While Bartle's taxonomy of players tries to segment player personalities based on the two orthogonal axes, the model that we are going to examine in this section tries to do so using no less than five. Accordingly, this framework is known as the Big Five Personality Traits taxonomy.

This model emerged out of research by Ernest Tupes and Raymond Christal, originally done in the late 1950s. However, it was popularized only in the late 1980s and 1990s by Digman (1990) and Goldberg (1993).

The big five personality traits postulated by this framework are labeled as *openmindedness* or openness to experiences, *conscientiousness*, *extraversion*, *agreeableness*, and *neuroticism*. Thus leading to the acronym OCEAN, under which this model is also known.

These five main traits are used as axes in the model, with extreme end of each axes described as follows:

- inventive/curious versus consistent/cautious for openmindedness,
- efficient/organized versus extravagant/careless for conscientiousness,
- outgoing/energetic versus solitary/reserved for extraversion,
- friendly/compassionate versus critical/rational for agreeableness,
- sensitive/nervous versus resilient/confident for neuroticism.

One important thing to mention regarding this model is that it was not developed as a result of any particular underlying psychological theory. Rather, it emerged from the observation of the experimental data. Several groups of researchers have noticed, after analyzing a large body of data gathered from personality surveys, that a large correlation exists between adjectives applied to the same person. Finally, those adjectives were grouped into the proposed five personality traits that we have mentioned.

Similar to Bartel's taxonomy and other psychological models, the profile of the person can be determined using the appropriate survey. Once again, the five dominant traits are treated as variables, with each person exhibiting a certain degree of each of them. The personality profile represents the position of a person in this five-dimensional space.

The proponents of this model claim that the degree to which each of these traits is exhibited is shaped by the early development of the personality and that they remain more or less stable throughout life. However, there are indications that individuals tend to become more agreeable and conscientious and less neurotic with age (Roberts et al. 2006).

Other factors that seem to have an influence on the big five personality tests include gender, the cultural origin of the person (Costa et al. 2001), and even the birth order. For example, in various tests, women seem to consistently score higher on neuroticism and agreeableness, while men, on average, score higher on assertiveness (Falk and Hermle 2018).

Furthermore, these traits are supposedly independent of the context and the situation in which a person might find himself, making them a good predictor of behavior in new situations. The big five have been applied to the analysis of academic achievements, learning styles, occupational fit, romantic relationships, and even political affinities. The big-five personality traits framework has been extensively applied to game design (Adams 2014; Dewanto and Tiatri 2021; Kachergis 2023).

The main advantage of the Big Five model over the other models we have discussed previously is its greater granularity, i.e., using five major axes as variables that define personality profile, allowing for a more nuanced representation than using just two. On the other hand, this is a general model, not designed specifically for the application to gaming. The biggest challenge with this model is to determine how a particular psychological profile reflects on gaming preferences and play patterns.

Player Personas

In the previous sections of this chapter, we have examined several approaches to segmenting the player base. As we have seen, each of these methods offers certain advantages and has its own disadvantages. On their own, each one of them offers only a partial representation of player profiles. Furthermore, they are, by definition, abstractions. The player-centric design philosophy implies designing a game with actual living and breathing human players in mind. It is hard to do so if you describe the person you are designing for as a set of numerical values; for example, she is a female, between 20 and 30 years old, with 5 on an openmindedness scale, 2 on conscientiousness, 1 on extraversion, 4 on extraversion and 1 on neuroticism,… oh and she is a killer ♣ according to Bertle's taxonomy.

Personas are intended as tools to bring a human face to this sort of description. They are meant to produce an actual concrete embodiment of a hypothetical person fitting one particular profile. This methodology was pioneered by Alan Cooper in the early 1980s (Goodwin 2009).

In a way, player personas offer a convenient way to synthesize and unify the profile data obtained by other methods. A typical way to construct a persona is to start with demographic data and add more personality to it. This is usually done by associating a suitable name and a photo that fits the desired demographic profile.

As persona is supposed to be an idealized representation of a whole demographic group, public web resources such as the Social Security Agency Register of Popular Names (SSA of US) and Unsplash and other collections of public domain photos can be of use (Unsplash n.d.).

The persona can be further expanded by including additional descriptions, such as the personality profile obtained by the OCEAN model of Bartle's taxonomy. Many people often include illustrative quotes to bring their personas to life. You can also augment your personas with any other source of data you might have at hand. Feel free to include the blood group, a horoscope sign, and the favorite flavor of the ice cream if you feel like it.

If you are building personas for an already existing game, the game-specific and feature-specific quantitative data can also be very helpful. For example, the female player from our previous example is called Emily, and she especially enjoys the real-time PvP mode in our hypothetical game, buys the Premium Pass within the first week of the season start, and spends about $5 per week on other IAPs.

If building a top-down persona for a potential new game, it is especially useful to make assumptions or include solid data on other games this type of player might be interested in.

One important thing when building personas is to keep a delicate balance between general and particular, if not peculiar. The more detail you add to a persona, the more descriptive it will be, but going over certain limits will create too idiosyncratic persona that will lose its general representative value. If you make your persona too quirky, it will cease to be an idealized representation of a segment of players and become a unique character of its own. Another risk you run is getting too attached to the details of a particular persona description. These are supposed to be helping tools meant to humanize the design process and not the actual people.

Finally, keep the number of personas low, especially if you are using them at the start of the development process. The ideal number should be between two and four. More than this, and you risk losing focus. Trying to make a game that is supposed to hit so many different targets is next to impossible. On the other hand, if you are building personas based on the existing data, the number of player clusters that emerge from the data should determine the number of needed personas. In this case, your game is already hitting all those targets.

There are several consulting companies offering player profiling solutions that combine all of these methods, most notably Solsten Inc. (Solsten n.d.).

In the most recent days, new AI-powered tools such as Large Language Models (LLMs), i.e., ChatGPT, have been suggested as potentially valuable in building personas (Butler and Gloor 2023). There are good reasons to believe that this is the case; LLMs are trained on an extremely large body of texts. By design, they provide generate their output using the relative weighted sums of word patterns, meaning they will, by definition, extract the most common descriptors of a particular person based on your prompt. This is supposed to ensure the generality of the produced profile. On the other hand, there are well-founded fears that the human touch would be lost in this automatized process. Although, I am inclined to think that the human touch remains within the handcrafted prompt.

Player Clustering via Data Analysis

With the exception of demographic data, all of the methods of target audience modeling that I have described in the previous sections of this chapter are qualitative in nature. They are all based on some sort of psychological model and rely in one way or another on self-reported data about player motives and behavior. These methods are especially useful at the start of the development process when no actual data is available.

However, there are several obvious limitations to these models. The first one is their reliance on self-reported data. There is a big difference between things that players think about themselves, the things that they are able to articulate, that they are willing to express publicly, and that they actually do. The role of rationalization and self-deception in decision-making is undisputed (Rorty and McLaughlin 1988). Secondly, these methods involve quite a lot of assumption-making. Finally, these methods describe who the players are, or at least who they are supposed to be, and not what they actually do within the game.

If you are working on an existing live game that operates as a service, you most likely already have a considerable amount of hard telemetry data. It makes sense to use this data to model your player audience. The typical approach to do so is to segment your player base using some of the so-called unsupervised machine learning methods (Duda et al. 2000).

There are many ways to do this. In what follows, I will describe one method that I encountered in practice and found particularly useful. This is not a data analysis book, and I will not go into the details of the implementation of this model and the methods that it uses. Rather, I will try to present a high-level overview of what it is supposed to do in order to demonstrate its key advantages.

In free-to-play games with rich and complex metagames. The behavior of engaged players typically involves a complex pattern of activities and interaction with several game features and aspects of the game. Players tend to cluster into groups sharing similar interests, motivations and consequently play styles. Trying to determine the interconnections between game features, their economy, and their influence on player behavior manually is very difficult. This method is meant to automate this process.

The method takes into account only the date for the players that have reached some critical point in the game. Typically, the criterion for inclusion of the data can be a player level when all major features are unlocked, i.e., the data of all players that have reached level 21 or a certain period of time that a player has been playing the game, for example, the data of player that have been playing the game for at least two weeks. The remaining data is disregarded.

Each player is represented by a vector of numeric values. The elements of the vector are values derived, such as the maximal level reached, LTV, amount of premium currency in the wallet, amounts of other currencies, average session length, average number of sessions, feature-specific progress indicators, etc. In general, the elements of the vector should include all high-level KPIs per individual player as well as any game-specific and feature-specific indicators that might be of interest.

In this way, each player is represented as a point in a multidimensional space. The dimensionality of the space is determined by the length of the vector. Individual values represent the coordinates of the specific player at this point. Your complete dataset will be represented as a $n \times m$ matrix K, where n is the number of players and m is the number of elements in each vector, i.e. the dimensionality of the representation space. Rows of the matrix K are vectors representing individual players.

The dimensionality of such representation is going to be too high for any practical analysis, but it is a good starting point for further processing of data. Furthermore, it doesn't provide obvious information about which of these variables are the most significant indicators of the player's behavior.

To make this data more manageable, a method called *Principal Component Analysis* (PCA) is applied (Jolliffe and Cadima 2016). The purpose of this step of the process is twofold. Firstly, the PCA method is designed to reduce the dimensionality of the data space. Secondly, as a consequence of this methodology, it extracts the interdependence between various features. This method has been applied in marketing research for a similar purpose for more than a decade (Desarbo et al. 2007).

What PCA does is create a new representation of data, a new $n \times m$ matrix $L = PK$. Here P is itself a $n \times n$ transform matrix. Thus, the matrix L consists of vectors, which are linear combinations of vectors taken from K.

In effect, each player is again represented by a newly constructed vector. The elements of this vector are weight values indicating the importance of variables in the original vector. Both vectors can be reordered into descending order using the absolute values of weights as a criterion. The first result of the process is the list of the dominant features, i.e., the elements of the original vector that are associated with the highest weight. These parameters have the highest impact on player behavior.

The space dimensionality can now be reduced by taking into account only a few first vector elements associated with the highest weights. The rest of the data can be ignored. Typically, either a fixed number of elements is taken, two, three, or four. The correct number can be chosen again by examining the weight values of the ordered weight vector.

Reducing the dimensionality of the data space allows us to now do the clusterization of players. Some standard, non-supervised clustering methods can be used, for example, K-means clustering, see (MacQueen 1967).

I will not go into detail how this method operates. Instead, I will just list its general steps. The K-means clustering begins by randomly placing k centroids into the representation space. The number of centroids k will determine the final number of clusters. There are various methods to determine the optimal number of centroids, see for example (Pelleg and Moore 2000). The distance of each vector into each centroid is calculated. The vector is assigned to the cluster defined by the closest centroid. Next, each centroid is moved to the means, i.e., center, of its own cluster. The process is then repeated until the convergence, i.e., until the change in the position of each centroid falls below a certain value.

As a result of this process, the player base is segmented into k distinct clusters. Each of these clusters defines one player profile. After the clusters have been extracted, it is possible to go back for the original data and do standard data analysis, but this time on a per-cluster basis. Things such as average spending, average session length, and average engagement time with particular game features can be extracted. This data can be further correlated with the qualitative data obtained by the methods we described earlier. Player personas can be augmented by this date, or the new personas can be constructed to represent the clusters.

Furthermore, the features that were selected as the principal components in the PCA step of the process are identified as the ones that best indicators of the player's behavior! This is very valuable information about play patterns and player preferences that can be taken into account when making design decisions.

Player Feedback

I still remember the launch of the first big feature that I designed. I haven't slept much the night before. I was tossing and turning in my bed, worrying about all the possible and impossible ways that things could go wrong. Most of all, I was worried about the players' reactions. Will they understand my design? Will they find it fun? Will they play it?

So many years later, and I still don't sleep before each update. For a game designer, the launch day is a bit like climbing on a rock stage, except that our audience can routinely be hundreds of thousands or even millions of people. It is our creation that we place in front of all these people. Every single one of these people will form an opinion about our work, at least to a degree, whether they like it enough to keep playing it or not. Others, a minority to be sure, but a vocal one, will form an opinion strong enough that they will feel the urge to express it.

For a game designer, opening a Reddit page or checking Twitter for feedback about his game can be a horror-inducing experience. The Internet is full of testimonies of game designers who got traumatized by player feedback. The audience can be harsh. People seem to be more willing to express negative emotions than positive ones. Maybe it is the zeitgeist, maybe it is our society, or maybe it is basic human nature. I am not going to lament over it.

Running a game as a service for six and a half years helps. Running a free-to-play game helps. It helps in many ways, but most importantly, it helps because it gathers a lot of data, enough of it to get to know our audience pretty well.

There are essentially two ways in which we gather player feedback:

- through telemetry data about player behavior
- via messages that players write about our game on social media, our own online forum, or send directly to our customer support.

"Your opinion is very valuable to us", is not just a corny phrase. We do actually deeply care about player feedback. We care about the quality of our own work and have a desire to do better, but we also care about the happiness of our players. Happy players stay longer with our service and spend more money. Player retention is one of our key performance indicators.

Those two ways in which we gather player feedback produce two fundamentally different types of feedback: *quantitative* and *qualitative*.

Quantitative feedback data is everything that can be measured directly by observing how players interact with the game. This includes things like engagement or the percentage of players that interact with a particular game feature, session length, retention, or the percentage of players that continue to engage with the feature after a specific number of days, etc. It can also include things that are more specific to a particular feature.

For example, the level that an average player has reached, the average score, the average amount of some virtual resource, etc.

Quantitative data is great. Numbers do not lie. They are objective measures of something. However, they do not tell the whole picture. They might tell you that something is happening within your game, but they will not necessarily tell you why it is happening.

Interpreting The Data

To get the full picture, one needs to take into account the other type of feedback, qualitative feedback, i.e., the stuff that players choose to tell about the game and their own experience playing it. As I mentioned, there is a multitude of ways players can express their opinion about the game, there are star ratings in app stores, official forums, customer support messages, official Facebook pages, unofficial Facebook groups, Reddit pages, and Discord servers, Youtube videos, and comments, of course, Twitter.

It might so happen that you release a new feature or an update for your game and that both of these two types of data go hand in hand. Players engage heavily with your feature, they spend time and money on it, and they are very happy about it, expressing their joy and gratitude on social media. Presumably, your customer support will not have much work in this scenario.

Well, this is all fine in theory, but in practice, things do not always go this way. You might also think that this is the only good scenario. However, this is also the only scenario in which you do not learn anything. All your basic assumptions about your design have been proven true; players are happy, and there is not much you should improve. Tinkering with the design further would probably cause damage.

But what do you do if things do not go this way? Of course, one possibility is that things will go horribly wrong and that both types of feedback will still be painting the same picture. The engagement is low and dropping, retention is terrible, players are raging on social media, and your customer support is flooded with complaint tickets. This might sound like a horrible turn of events. To be sure, pissing off your player base is never a good thing. However, the silver lining here is that the combination of quantitative and qualitative data will most likely help you pinpoint and diagnose the problem. Unless you are Ed Wood of gaming, you should be able to form an action plan and fix things somehow.

The third scenario is the most difficult and the most interesting one. It is everything in between. It is a scenario in which two types of feedback paint a very different picture. It has several subtypes that can have a very different effect on the game.

It often happens that your players seem really happy with the content or the feature you have just released. They will write glorious reviews on Reddit and share screenshots on Facebook, etc., etc. Yet the qualitative data will tell you that your feature is just not performing, the engagement might be low, retention might be dropping, or monetization will be lacking. This might seem like a paradox. If they are happy, why isn't this obvious from data?

The answer to this is simple: only a tiny minority of people will care enough about your work to voice their opinions. All qualitative data that you can get originates with this vocal minority. The silent majority of players that generate the bulk of qualitative data do not share the same impression. Fixing this situation can be extremely difficult. In order to lift your key performance indicator numbers, you might be required to alter exactly the things that the vocal minority likes. They might react negatively only to reinforce the already negative feelings of the silent majority, pushing you down the spiral of doom. Therefore, always keep in mind that qualitative feedback originates with the vocal minority, and qualitative data originates with the silent majority!

There is, of course, a vice-versa scenario. Your players are crying bloody murder, yet all your KPIs are going through the roof. This obviously seems like another paradox. If they hate it, why do they keep playing it? Why aren't our quantitative indicators tanking? Should we do something about it? Should we panic?

The key to understanding this situation is human nature. What we are dealing with here are actually three quite distinct things:

1. things that people are willing to say aloud,
2. things that people are willing to admit to themselves,
3. things that people are actually thinking, feeling, or doing.

We, humans, are not particularly good at articulating our own emotions. Quite the contrary, we are quite good at making a mess of them. When facing something new, some people will express negative emotions even if they are actually impressed.

Some people are just like my mother-in-law. For them, expressing negative emotions means that they are actually impressed by something. For others, negative emotions might mean that they see the change as a new gameplay challenge. It means that they see what you did, see a new challenge, and are stepping up to it. This is especially true if your new feature is forcing veteran players to rethink their strategies and relearn the skills that they have already acquired. This might sound scary, but it is actually a very good thing! Just like muscle aches after intense exercise, this is a sign of new growth.

If you are facing this scenario, my advice is to sit down and relax. Let the dust of the launch day settle. Both the opinions of players and telemetry data will evolve in a matter of days. Most likely, the two types of feedback will start to converge into something actionable. Making rushed decisions in this scenario can lead you into an even deeper mess. Keep in mind that people do one thing, say another thing to themselves, and say aloud yet a completely different thing.

The last scenario that I want to mention is the one that I personally dread the most. It is a scenario of mediocre emotions and mediocre or non-existent impact on quantitative data. Imagine a situation in which you have been working hard on the content or a new feature for several months. The launch day finally arrives, but all you get from your players is a big giant "meh"!? Your effort has essentially been wasted. There is nothing to fix cos there is nothing that needs to be fixed. Players kinda appreciate your effort but are not really falling head over heels for it. There is nothing broken in your system, it is just not whooming with enough oumph.

It is not so much about if emotions are positive or negative; it is about the strength of emotions. It is about the amplitude, not the direction!

In my personal experience, it is not about the direction of the emotion but about the amplitude. It doesn't matter if players are angry or happy about something as long as they feel emotions strong enough to wish to express them loudly. If you get an intense emotional response, and it shows in your telemetry data, it means that you are onto something!

Understanding Time

5

Modeling Player's Time

There are many ways one can think about games. In his popular book on game design, "The Art of Game Design: A Book of Lenses," Jess Schell (Schell, 2008) offers a multitude of lenses through which you can examine your designs. However, when you are designing a mobile game or any game that would run as a service, you are designing for player retention. A very useful way of thinking when faced with such a task is to think in terms of time. Essentially, what you are doing is structuring the player's time.

Ideally, the game that you are designing will keep players transfixed for months or even years. Still, even a journey of a thousand miles starts with a single step. All that time will be constituted of a myriad of individual play sessions. In this text, I will focus on the anatomy of these atoms of playtime. Understanding their structure can help us design better games.

The designers of console and PC games have it easy. Their players usually plan their own time and set aside hours dedicated to their gaming habits. They know that they are going to play Apex Legend with their buddies that evening and that they are going to be playing Ghost of Tsushima on the weekend. Of course, they do. Planning a multiplayer session with three other guys requires coordinating everyone's schedule and they just paid sixty bucks for that new AAA game. Their gaming time is structured around the titles their attention is focused on.

In the world of mobile, players rarely premeditate and plan their playtime. Mobile games are here to fill in the gaps of free time that every one of us will have during a typical day. You are waiting in a queue, you are stuck in traffic, you need a quick coffee break… There are a couple of minutes you want to fill with something more entertaining than staring through the window. You pop up your phone and start browsing its home screen, looking for a game that would do the job. Any that you already have installed

© The Author(s), under exclusive license to Springer Nature Switzerland AG 2024
S. Stanković, *Game Design for Free-to-Play Live Service*, Synthesis Lectures on Image, Video, and Multimedia Processing, https://doi.org/10.1007/978-3-031-56156-6_5

will do, as long as a meaningful gameplay session can be squeezed into this small gap of available time.

These two halves of the gaming world have two radically different play patterns when it comes to time engagement. Typically, PC and console games are enjoyed in one or two play sessions per day. These play sessions will usually last from at least 30 min to several hours. Most people can reserve so large blocks of free time only at certain times of the day, usually in the evening or at night. On the other hand, mobile games will be played in several bursts of gameplay dispersed throughout the day. These sessions can last anywhere from under a minute to up to as much as 15 min. Depending on the title, these short sessions can be mixed with longer sessions, again usually the time of the day when leisure time is more plentiful for most people, i.e. in the evening.

During working hours, your game will need to compete with attention with the player's actual daily duties and with all sorts of other procrastination-inducing media, such as social networks, Instagram and Twitter, news portals, etc. In the evening hours, your competition will be console and PC games, movies and TV shows, Netflix and Hulu, etc.

If you are designing a mobile game, your game needs to, like a vine, grow its way into every free nook and cranny of a player's daily schedule. In order to do so, it needs to offer play sessions of a very flexible length capable of both fitting into narrow gaps of free time during the day but also engaging enough to fill in the longer time periods!

Remember, Every moment of free time that a player is not playing your game, he is going to spend playing someone else's game!

Player's Journey

Contemporary games, and especially Free-to-Play games, can be exceedingly complex with myriad features, gameplay modes, and loads of content of different types. When designing a game, it is easy to fall into the trap of thinking about it in terms of individual features and focus on the design of particular aspects of the game. However, this is not how the players see the game. They do not see it as a set of individual components. They tend to experience it as one holistic piece of entertainment. Instead of throwing various things at the player in a piecemeal fashion, as a designer, you should strive to design one unified user experience in which various features and types of content complement each other, creating synergy.

One method that can be especially useful in this endeavor is mapping the *player's journey*. This term denotes the path that the player is expected to take through the game. The player's journey includes topics such as the player's onboarding, the introduction of the core gameplay, the transition into the metagame, progression through the metagame, and any possible emotional and narrative arcs that the game might have.

Different types of games pose different challenges when it comes to mapping the player's journey. Arguably, single-player premium games offer the simplest case of this design task. The designers, in general, have a great deal of control over the path and the speed that the player can take through the game. Other types of games, such as open-world games, sandbox games, various simulation games, etc., pose much greater challenges. By their nature, these games allow for very diverse playstyles and offer multiple paths that a player can take through the game. This is also true for a lot of types of Free-to-Play games that are operated as a service. Furthermore, these games typically evolve over time. As a result, the map of the player's journey in such games tends to be much looser than in the case of single-player premium games.

When modeling the player's journey of the Free-to-Play games, or in general, any GaaS, it is important to keep in mind that the longer the time scale that we are operating on, the less precise the description will be. There are two big reasons behind this phenomenon. In Free-to-Play games, one of the key factors influencing the speed by which the player progresses through the game is the amount of time that a player is willing to devote to playing the game during the day. Highly engaged players will simply sink more hours into the game during any given day. They will have more play sessions in the day and spend more minutes in each play session. Consequently, they will make their way through the game much faster than a casual player will. This is compensated to a certain degree if a game is using some strict sessioning mechanism, for example, an energy mechanic. However, even in such games, players can circumvent these restrictions by spending money. Things are much more complex in games that do not employ such methods. In games that have a complex metagame, as new features and gameplay modes unlock, the game allows more and more divergent playstyles to evolve, thus making mapping the player's journey much more difficult. The prime examples of this are farm simulations and the city builder games.

The concept of the player's journey shares many similarities with the customer's journey, a well-established concept in marketing and product design. When designing the customer's journey, marketers divide it into several discrete phases. This is paralleled in the world of game design. When talking about Free-to-Play games, the game designers and product managers often talk about the early gameplay, mid-gameplay, and the elder game. This indicates the three phases of the evolution of the player's relationship with the game, i.e., the three big stages in the player's journey, as shown in Fig. 5.1.

The *early gameplay* includes the period of the journey in which the player is getting introduced to the basics of the gameplay, usually to the core gameplay and to the core pillars of the metagame. This phase of the game includes what is known as the *First Time User Experience* (FTUE). The accent in this phase of the player's journey should be on the player's *engagement*.

The *mid-gameplay* is part of the player's journey when the player is already familiar with the game and is gradually unlocking new features. During this phase of the journey, the player's attention should move from the core gameplay to the metagame. The different playstyles evolve during this phase of the game. The focus moves from engagement to the *retention* of the player.

Fig. 5.1 Phases of the Player's Journey

Finally, in the elder game phase of the player's journey, the player is expected to be very familiar and already greatly emotionally invested in the game. The players that make it to this phase of the journey are already veterans; quite often, they are the game's superfans. Usually, they already have their distinct set of habits within the game, ensuring their engagement. They are expected to focus only on a subset of the metagame features. As these players obviously retain quite well, the focus in this phase on the player's journey is usually on the *monetization* of the game.

When we mention engagement, retention, and monetization in this way and tie them to specific parts of the player's journey, this doesn't mean that we should not pay attention to them in other parts of the journey. It simply means that the accent should be on them at a particular phase. If there is no engagement, there can be no retention, for example. If a player has been playing a game already for several months, it means that the game is already able to retain this player.

The distinction between FTUE and what happens beyond FTUE is especially important and should be especially underlined. We are going to discuss the design of the FTUE in detail in one of the following sections of this Chapter. But it is important to note that the FTUE is typically a much more tightly controlled part of the game's user experience than the rest of it. In terms of the player's journey, mapping the FTUE usually means mapping the order in which the player goes through its various steps.

Mapping the game beyond the FTUE means planning the order in which the player is expected to encounter and unlock various additional game features, including the new gameplay modes and the types of content. It also includes trying to predict the velocity with which the player will move along the *various progression* vectors and consume the content of the game. We are going to talk more about progression vectors in one of the following sections of this chapter.

Consuming the content of a game can mean several things depending on the type of game we are making. In a heavily level-based game such as for example various Match-3, bubble shooter, or other causal games, this could mean the speed with which the player moves through the levels. On the other hand, in Card Collection Games or Character Collection Games, this could mean the velocity by which the player unlocks and upgrades individual characters or cards. It could also mean the speed by which the player unlocks various chapters in a narrative-driven game. In many games, it means a combination of several of these things.

As we have already mentioned, when we are designing a player experience, we are actually designing players time. Therefore, when we are mapping out the player's journey, we are mapping out the time and the rhythm in which the player engages with the game. There are two fundamentally different ways of thinking about a player's time.

One obvious way is to think in terms of real-time units, i.e., minutes, hours, days, weeks, and months. In contrast, the other way is to see time as a series of discrete events, usually distinct important events that happen in the game, such as a player completing a step in the game tutorial, or unlocking a new gameplay mode, or defeating a boss. Both of these approaches have distinct advantages. The first approach includes mapping player time within the game but also outside of the game, i.e., breaks between play sessions. It doesn't necessarily describe the sequence in which the player is expected to take certain actions. The other approach takes account of the sequence of events but typically doesn't include the exact time intervals in which these events take place. It also ignores the time that the player spends away from the game. Fully mapping the player's journey actually requires the use of both of these methods. Each of these methods has its advantages when applied to specific parts of the player's journey.

Typically, the player is expected to go through the bulk of the FTUE within the first session of the gameplay. Among other things, the FTUE usually includes at least some sort of a game intro and a tutorial. Individual steps of the FTUE are of great importance. Therefore, the player's journey through the FTUE is typically modeled in terms of these discrete steps. Keep in mind that the FTUE of any game, especially a Free-to-Play game, begins well before the player actually installs the application on a device. Furthermore, bits of FTUE can take place well after the first gameplay session. Individual gameplay features can have their own FTUEs.

Moving beyond the FTUE, the rest of the game is better modeled in terms of real-time units. Start off by creating the order in which the player is expected to unlock various features of the game, then plan where a player should be at a particularly important time milestone, i.e., after one day of gameplay, after three days, at the end of the first week, after two weeks, after one month, etc. The important milestones of the player's journey can match the typical retention metrics such as D1, D3, D7, D30, etc. Think as far ahead as possible. Think about what is realistic and what is not. We are going to take a look at the practical examples of this in Chap. 6 of this book when we talk about how to think when balancing a game.

Roadmap and Production Issues

Mapping out the player's journey is important for the design of the player experience. However, it is also important for several other reasons. When designing a new game, it can help plan the development schedule and organize the operations of the live service. When launching a new game, the first cohort of players will begin their player journey with the start of the game's live service. It means that they will hit certain player's journey milestones in weeks, months, or even years. This gives plenty of time for the team to develop and implement these features. A GaaS game does not need to launch with all the game features. You can use the player journey to plan out the minimal set of features that the game needs to have at the soft launch, at the global launch, and even at the end of year one of the live service. Indeed, quite a few games launch with the entire elder game missing. Sometimes, this is a conscious decision by the development team. Sometimes, this is a result of necessity. Focusing the team's resources on the part of the game that the players will not encounter for several months doesn't really make sense. On the other hand, many teams do not actually yet have a clear idea of what a suitable and profitable elder game should look like. This is fair, and by itself, it is not necessarily a sign of trouble. A Free-to-Play game needs to find its audience and needs to be able to evolve after the launch.

If you are working on a live game, the player's journey mapping can help you plan your release roadmap. It can help you design appropriate new gameplay features. Some of these features can be aimed at players at the specific phase of the player's journey. For example, you might decide to create a new gameplay mode aimed to lift engagement during the mid-gameplay, or you might create a new retention feature for the elder players. When designing features that straddle several phases of the player's journey, it is especially important to think about how they will look to players in each of the three big stages: early, midgame, and elder players.

These three big groups of players are also important to keep in mind when designing the scope of each new big game update. It is OK if the update is organized around a new feature or new content aimed at one of these big player groups. However, the other two groups should not be neglected and excluded. Be sure to include bits of content that each of them can enjoy. Neglecting to do so is a surprisingly common pitfall. Quite a few, even experienced teams, tend to focus on the most prominent and usually the loudest part of their player base. These are the emotionally most invested of the players. Inevitably, these tend to be elder players. There is nothing more demoralizing for the early players than a big game update full of content that you will not be able to unlock for months in advance. In the long term, this inevitably leads to a drop in player retention and can be detrimental to your live service.

Mapping out the expected rhythm of content consumption is important for other reasons, above all, for production and operational reasons. The love service of the game needs to be operated in a sustainable fashion. If regular updates of content are expected, the speed by which the player is expected to consume it must be regulated in order to ensure that players will remain engaged with the previous batch of content until the next batch can be produced and delivered. In general, the content treadmill is unsustainable if it takes more time for the team to produce the content than it takes the players to consume it. If this is the case with your game, you need to rethink your metagame and the production pipeline.

Progression Feedback

In the previous section, I mentioned a term that I haven't actually defined. This was a deliberate choice on my part because this term is important enough to warrant its own section. The term in question is a *progression vector*. This term is not very common, but I find it very useful as an umbrella term encompassing a whole range of diverse game features and design tropes serving the same function of providing the *progression feedback* to the player.

As we have discussed in the previous section, the player's journey is an expected path that the player takes through the game. This journey is a result of the player's interaction with the game. Whenever the player manages to achieve something within the game, the player receives some sort of feedback about this. A victory screen might pop up, a reward might be given, a score counter goes up, a status bar gets filled, etc.

If such achievements happen in a sequence, for example, if the player gets a new reward whenever he reaches the next new score milestone, this constitutes a progression vector. The progression vector is a persistent indicator of player progress through the game. In other words, a progression vector is any sequence of milestones that can be used to indicate the player's progress through the game or one of its features. To advance along the progression vector, the player is required to achieve something in some sort of regular intervals. The progress vectors provide feedback about the player's successes in a systematic fashion.

A broad range of very standard game features serves this purpose. An *overworld map* serves this purpose in classical platformer games, such as Super Bario Bros. In level-based games, casual games such as Candy Crush Saga a *saga map*, can be used for this role. It is a winding path connecting dots that indicate individual levels. The player needs to pass levels in the predefined sequence. The level that the player is currently playing is indicated, providing a visual representation of the player's progress through the game. In a similar way, a leaderboard or a ladder of leagues or arenas serves this purpose in many games. Figure 5.2 shows the level selection UI of the Angry Birds Classic game, which serves the same purpose.

The progression vector does not need to be visualized as a path or a line. What is important is that the player's goals are arranged in some sort of a clearly understandable sequence. Any system of *experience points* and *experience levels* constitutes a progression vector. In this sense, even features such as *premium passes* work as progression vectors.

Furthermore, any feature that indicates gradual progression serves this purpose. In many character-based games, *character collections* and galleries are the main progression

Fig. 5.2   Angry Birds level selection UI

vectors. Any sort of *collection album* consisting of a series of items that can be collected or unlocked is also a progression vector. The City Album of SimCity BuildIt is shown in Fig. 5.3. Finally, even the *technology trees* and similar features act as progression vectors in some games.

Any game that has even a rudimentary game has at least one progression vector. Many, if not most, games employ more than one progression vector. In general, the richer the metagame, the more progression vectors we can expect to encounter. Some of these progression vectors persist throughout the game, while others can be tied to particular features or even exist only for a limited amount of time.

Fig. 5.3   The City Album feature of the SimCity Buildit acts as a progression vector

Fig. 5.4 Progression vectors in SimCity BuildIt, city level, war rank, and mayor's pass

For example, SimCity BuildIt employs a myriad of progression vectors. The player's city level, based on the XP score and the population of the player's city, are the persistent progression vectors present throughout the game. On the other hand, the player's war rank is tied specifically to the War of Disasters feature and depends on the player's engagement with this feature alone. Finally, the Mayor's Pass is a progress vector that exists only for the duration of one Mayor's Pass season. Figure 5.4 shows these three progression vectors in SimCity BuildIt.

The reason behind this is to avoid a situation in which a player gets stuck in progression along a single progression vector. The progression vectors are an important tool for player retention if the player sees that his progress is blocked beyond hope of progress, this typically leads to the player abandoning the game.

The diversity of these features can easily lead us to the conclusion that there are nearly an infinite number of ways by which a player can advance along these vectors. To a degree, this is true. There is nearly an infinite number of actions that a player can be rewarded for. However, there are, in essence, only four ways in which a player can advance along any type of progression vector in any game: skill, effort, time, and luck.

For example, in the classic Angry Birds game, a player needs to beat each level in a sequence using his slingshotting skills. The same applies to climbing through arenas in Clash Royale. On the other hand, the player's ability to advance along the saga map in many Match-3 games depends more on luck than on the player's skill. In idle games, such as Idle Capitalist or any sort of cookie-clicker games, a player's progress is simply a function of time. Finally, very many Free-to-Play games reward players for engagement and effort rather than for success. The examples for this can range from minimal engagement, i.e., simply logging in to a game daily rewards, to giving achievements for constantly repeating a certain action. Mastery rewards in Hay Day are an example of this. The player is rewarded with a certain amount of machine experience points whenever a

Fig. 5.6   Mastery levels in Hay Day

particular crafting machine is used. If the player collects enough points, a new mastery level is awarded, as shown in Fig. 5.6.

Each of these strategies on its own is not perfect. Skill-based progress vectors are bound by the player's learning curve. This is a topic that we are going to address in a separate section later in this chapter. According to the SDT framework that we discussed in Chap. 3, luck-based progression vectors have a serious disadvantage in that they do not satisfy the player's need for autonomy. The same applies to purely time-based progression vectors. Rewarding a player's effort is arguably the safest strategy in Free-to-Play design.

To avoid the individual pitfalls of these strategies, many games employ a combination of multiple progression vectors employing a combination of reward strategies.

In addition to providing the indicator of the player's journey through the game, progression vectors are typically used as unlocking criteria for various game features and bits of content. For example, Fig. 5.7 shows the trophy road in Zooba, used as an unlocking method for additional characters and content.

Finally, in many Free-to-Play games, a player is able to use real money to advance along one or more progression vectors. Typically, this is done indirectly by purchasing items within the game that make gameplay in some way easier or allow skipping of time. This monetization strategy is known as *selling progression.*

Fig. 5.7 Example of the trophy road as a tool for unlocking new content in Zooba

Player's Goals

One way of thinking about games is by imagining a game as a set of goals that the player desires to achieve and obstacles preventing him from doing so. It is through overcoming these obstacles and reaching the goals that the player derives satisfaction from playing a game.

The goals that the player is trying to reach can take a myriad of forms. They can be anything, ranging from banal, like jumping over a gap between platforms to divine, defeating a mega-powerful hell-spawned demon; from abstract, like eating that pixel dot, to highly complex, like optimizing the resource flow of a virtual civilization. The form in which goals are presented to players is dictated by narrative tropes of the game genre. The goals in games are as diverse as the human imagination.

However, on the system level, all these goals share some fundamental similarities. We can speak about player goals in terms of the time and effort that the player needs to invest in order to reach a particular goal. Generally speaking, the required effort is determined by the difficulty of obstacles that we put in front of the player and his skill. Balancing these two variables is in itself a huge topic in game design.

In this section, I am going to focus on the second variable, the time. I believe that this is both important and useful because one of the key metrics used to evaluate the success of the sort of games I make is retention, which is obviously defined as a function of time. Essentially, when we are optimizing a game for player retention, what we are doing is actually structuring the player's time by influencing his aspirations.

In terms of the required time, the goals can be divided into three broad categories:

- Short-term goals
- Medium-term goals
- Long-term goals.

Short-term goals are what is usually referred to as moment-to-moment or minute-to-minute interaction. This is a domain of the core gameplay.

On the opposite side of the spectrum, long-term goals are something known as aspirational goals, something that a player tries to achieve but is by no means guaranteed to reach. Indeed, according to some estimates, even in the world of AAA games, on average, only about 10% of players ever reach the end of a game.

From the retention point of view, the medium-term goals arguably have the biggest importance. These types of goals form the bridge between the moment-to-moment gameplay and the ultimate aspiration goal. In a way, they form the backbone of the retention structure, and the pursuit of medium-term goals will occupy the bulk of the player's time. This is what is usually referred to as the metagame.

Although the timescales of all these types of goals grow seamlessly from the shortest to the longest, there is one curious thing that must be noted. In most games, these goals are presented in a peculiar order.

Typically, the player is presented with an aspirational goal, the ultimate prize, the longest-term goal that the game has to offer. This establishes the context of the game and the world that the player is about to enter.

Next, the player is presented with a short-term goal. In most games, a player would go through some sort of tutorial that would teach him about the basics of moment-to-moment interaction.

Finally, and only after the player has established at least a rudimentary level of skill, medium-term goals are gradually revealed.

To better illustrate what I am talking about, I am going to use examples from two games that I hope almost everyone should be familiar with:

- Super Mario Bros
- Football (European version, known as Soccer in the US).

Example 1: Super Mario Bros

The story of every game in Nintendo's Super Mario Bros franchise is somewhat psychedelic. You are an Italian plumber. A giant turtle monster has stolen your girlfriend and you are expected to rush through a kingdom inhabited by moving mushrooms to save her.

The intro scene presents us with the aspiration goal: save Princess Peach!

Immediately after the end of the cutscene, you are thrown into action. The short-term goals become obvious. You are supposed to jump over the gaps between platforms and eliminate the malicious goombas.

Your struggle is not endless. Eventually, you will reach the end of the level (or a halfway checkpoint). The first medium-term goal is revealed. You finally reach the flag-pole that marks the end of the level. Victoriously you walk into the castle, just to face the disappointment: the princess is in another castle! In the very next moment, you emerge to the overworld, a map of sorts that reveals even more medium-term goals. The game reveals its complexity to you.

Therefore, when it comes to goal structure, the game of Super Mario Bros can be summarised as follows:

- Long-term goal—save the princess
- Short-term goal—don't die, jump over gaps, kill enemy mushrooms
- Medium-term goals—reach the end of the level, clear all the levels in the current world.

Example 2: Football

Ostensibly, football (known as soccer in North America) is a game of kicking the ball using only your feet and legs. It is safe to say that not one of the millions of kids who started to play football with any level of seriousness was thinking about it in these terms. Of course not—their dream is to become champions. Champions of the world, champions of the continent, of the league, of whatever. Becoming a champion is the ultimate aspirational goal in the world of football.

Obviously, to do so, one must beat opponents. Minute-to-minute gameplay of football is about dominating the playfield and scoring goals. Scoring more goals than the opposing team will result in winning a match!

Winning a single match is rarely enough. The team must win many matches to climb the league table or reach the tournament finals. After a single league season ends, the next one continues. Winning the domestic championship will let the team qualify for international competition, etc.

The metagame of football developed over more than a century and is now extremely rich, consisting of layers upon layers of medium-term goals.

Short-term goals: score a goal, win a match,

Medium-term goals: win a match, climb the league ladder, win the season championship, qualify for the World Cup, etc.

Long-term goals: win the continental Cup, win the World Cup, win the World Cup five times in a row, etc.

Another important thing that the metagame of football illustrates well is that the timescale of goals can shift depending on the situation. What a medium-term goal is in one context can be seen as a short-term goal in another. Winning a match can be seen as a medium-term goal in the context of a single round of competition, but it can also be seen as a short-term goal in the context of the whole season. Furthermore, new layers of goals can be stacked on top of existing ones. Even for the teams that reach the ultimate prize, like winning the World Cup, the goals can shift. The objective becomes winning a world cup twice, winning it twice in a row, winning it five times, etc.

Finally, I want to underline some key takeaways from this section. A game can be seen as a set of goals and obstacles that the player needs to overcome to reach these goals. Goals can be very diverse but ultimately can be categorized according to the time that a player needs to invest in order to reach them. By constructing the game in this way, we are structuring the player's time.

Short-term goals are part of core gameplay and moment-to-moment interaction. Long-term goals are aspirational goals. Medium-term goals form the bulk of the metagame. Goals are typically presented starting from the aspirational goals, then short-term immediate goals, to medium-term goals.

Density of Goals

The notion of *Density of Goals* is so important for a successful metagame design that I decided to devote a separate section to it, although it might just as well have been a part of the previous section in which we have already talked about the importance of goal formation. We also mentioned that not all goals are created equal and that they can differ in their importance, time horizons, and difficulty. We made a distinction between long-term aspirational goals, short-term, minute-to-minute goals, and medium-term goals that make up the bulk of the gameplay.

However, in that discussion, we used a pretty simplistic model in which goals are presented in a sequence. In this model, a player still needs to clear goals one after another. In order to reach the next goal, the player needs to overcome every obstacle in his path. This type of gameplay feels very familiar and intuitive. After all, this is the rough structure that so many well-known games have, starting from old-school arcade adventures and platformers to many more recent games such as Angry Birds or Candy Crush (Fig. 5.8).

This type of structure, known as the Linear metagame, has some obvious advantages. Above all, it is very simple to design. The game designer controls every step of the path that the player must take through the game. He can precisely control the difficulty of the challenges that the player will face. In addition, this structure is very convenient in the narrative sense. The player's progress through the game can be easily accompanied by a set of appropriate cutscenes that tell the story and introduce the new gameplay elements.

Fig. 5.8 Cave Story, Owlboy, Bioshock Infinite—No straying from the straight and narrow path!

However, this model has its own Achilles' heel, an inherent weakness that makes it especially unsuitable for certain types of games, namely games employing the free-to-play monetization model. The problem itself is the one described by rhetorical questions from the intro of this text. Namely, it is very likely that a player will eventually get stuck on a particularly difficult point in the game. His path forward can get blocked by a particularly nasty opponent or a difficult level. Indeed, only 40% of players complete a typical single-player game with a linear structure.

If you are a game creator, this fact is not necessarily a problem in the world of premium games. Your customer has already paid the full price of the game, and you do not have to worry if he has reached the end of the game or not. As long as the player feels satisfied with the part of the game he managed to play through, you can count on him buying the sequel.

Things are radically different if you are running the game as a service, for example, a free-to-play title. In this world, player retention is the key to financial viability and success. Player churning because he got stuck on a level is a Product Manager's nightmare!

Non-linear Gameplay

The need to avoid this glaring problem has caused the rise of various other metagame structures, all known under the umbrella name of Non-linear gameplay. The obvious idea here is to provide the player with an alternative path of progress if he has become blocked by insurmountable obstacles on the main path.

This approach evolved relatively early in the history of video games, and it can be seen, for example, in overworld maps in Super Mario World on SNES, where the player has a limited choice over the order in which he can attempt to traverse particular levels (Fig. 5.9).

But what happens if the player gets blocked on the alternative path as well? Obviously, we can always add more paths and more goals for him to go after. This brings us to the real topic of this text: the concept of the density of goals.

Fig. 5.9 The Cuphead world map offers a limited choice of paths that the player can take

The idea behind this concept is that the game should be ready to offer a variety of goals at every step of the player's journey. The player can make his own choice about which goal to pursue at what time. This choice can be driven by his skill, his play style, or simply his mood on a particular day.

The most important feature of this approach to the metagame design is that it dramatically reduces the possibility that the player would get hopelessly stuck at any point in the game. Even if his previous goal turns out to be for the moment unreachable, he is free to pick and choose another one that might be within his reach.

The density of goals dramatically reduces the possibility that the player would get hopelessly stuck at any point in the game.

RPG Mechanics

Creating such a structure with traditional methods can be difficult, especially if one is still focused on presenting a coherent narrative in the game. Enter the RPG mechanics! The letters in this abbreviation stand for Role Playing Games, the type of games where this metastructure originated. However, nowadays, this game design pattern is applied in a variety of genres, some of which can be quite distant from the original.

The RPG mechanics push the idea of goal density to the extreme. The focus in these games is not on the narrative but on a character and its development. In more standard, more linear games, characters have permanent qualities or qualities that evolve little during the game. In contrast, characters in games employing RPG mechanics are described by a set of numerical values indicating their properties. For example, a typical character could have variables describing his properties (Fig. 5.10).

- Strength—the amount of damage he can inflict on the opponent characters,

Fig. 5.10 In Genshin Impact, characters have a multitude of properties divided into several categories!

- Health—the amount of damage he can take before being vanquished by the opponent,
- Speed—maximal velocity at which he can travel,
- Stamina—maximal time during which he can run at a high speed.

Each of the character properties can be upgraded in some way. RPG games provide a variety of sometimes very elaborate methods of upgrading mechanisms. The player's goals become about upgrading each of these values to the next level. The attention shifts from the progression through the story to the development of the character. This can in itself provide a very high density of goals.

However, it can be multiplied several times by the introduction of a multitude of characters at the player's disposal. The player's game starts to revolve around upgrading each of the character properties for every character in his roster. Expanding the character roster by unlocking additional characters becomes another important source of player goals.

The player can, thus, move forward through a game along a series of parallel progression vectors.

This type of metagame is actually quite versatile. It has been adapted to a variety of game genres on almost every platform, from very hardcore games, such as Genshin Impact and Summoners War, to more casual titles, such as Brawl Stars (Fig. 5.11).

The basic RPG model has spawned additional metagame patterns. The so-called Card Collecting Games (CCGs) are an example of this evolution. Here, the focus is on the endlessly expanding inventory of cards representing a broad roster of characters. These

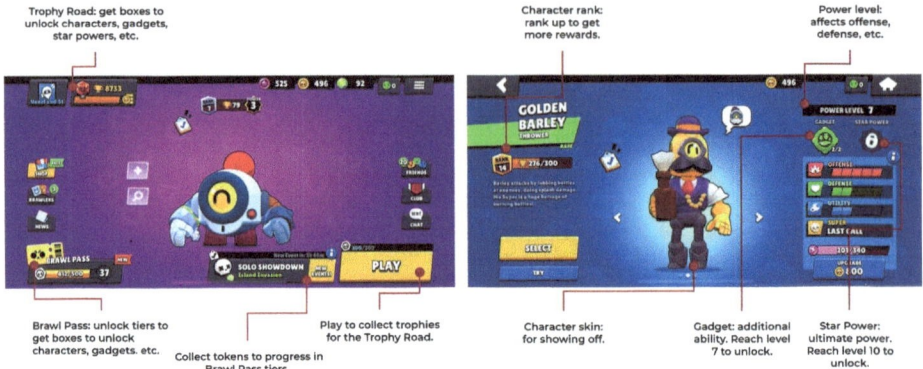

Fig. 5.11 Brawl Stars: density of goals in the character metagame

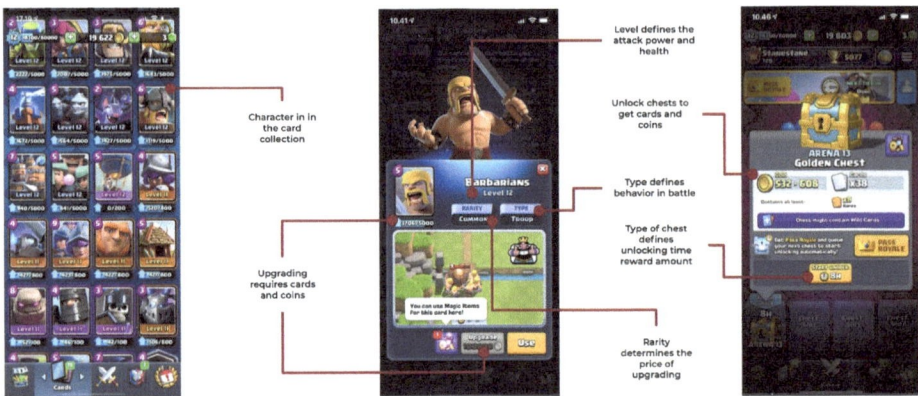

Fig. 5.12 Card Collection metagame in Clash Royale

games employ a somewhat simplified, very intuitive metaphor of collecting cards as a mechanism for unlocking new characters and upgrading the existing characters (Fig. 5.12).

We are going to return to RPG mechanics in a subsequent section of this chapter to examine them from a slightly different yet related perspective of the player's learning curve.

Sandbox Games

While RPG and CCG games offer a great density of goals and present the player with a smorgasbord of choices, they still share one key property with linear games. All goals

Fig. 5.13 Examples of various things a player can do in Minecraft

that a player can pursue are defined by the game system. However, there is a type of game that does away with even that constraint.

"We are building a McDonald's restaurant now. We just killed the Ender Dragon. Do you want to see the Sakura tree that I made? I used Pink Wool for the leaves!" is an honest—to—god quote of my 8-year-old son during the Minecraft gameplay session.

"It's so lovely to see him play so nicely, but I have no idea what is the goal of this game!"—Honest-to-god quote from my 71-year-old father about the same Minecraft gameplay session.

Only in Minecraft do you get to kill a dragon, build a lighthouse, or construct complex machinery in the same session (Fig. 5.13).

Instead of a predefined metagame structure, sandbox games employ a set of rules that define the way in which the gameplay elements can interact. Akin to atoms in the real world or LEGO bricks, Minecraft blocks can be combined in an almost infinite number of ways. This robust yet loose game system allows for what is called emergent gameplay. Precisely, because the game is left to interpretation, its world allows for players to formulate their own goals. These goals can shift, be transformed, or be abandoned in an instant.

Let's build a treehouse! Let's go dig for diamonds! Let's make a SpongeBob sculpture or a Redstone roller coaster!

The density of goals is infinite, as is the feeling of autonomy that the player feels. It is impossible to get stuck in Minecraft!

Learning Curve

Have you ever started a new hobby only to drop it after a few weeks? You might have paid a membership fee to a club, bought fancy equipment, and even paid for classes, only to give up after a couple of months. It might have been pottery classes, it might have been golf, it might have been capoeira or painting or trying to learn Japanese. No matter what it was, the journey probably looked like this (Brassey et al., 2019).

The first time you tried it, it was a bit awkward. You were not good at it, but it is to be expected. You are there to learn something new. You were pumped with enthusiasm,

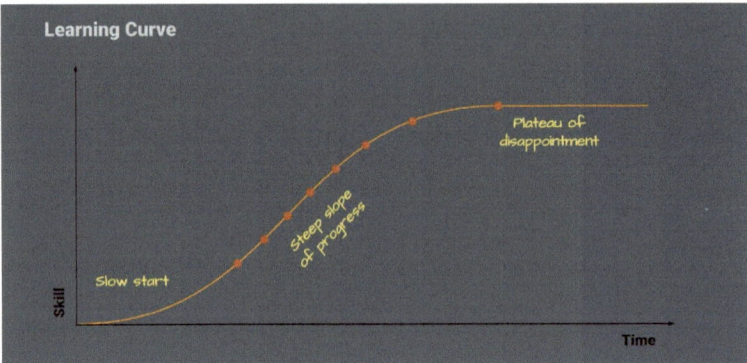

Fig. 5.14 A diagram of a learning curve

and the sheer joy of trying on something novel was exhilarating. The next few times you went, it was way better. You could really feel your skills improving, and the improvement was dramatic! You really enjoyed yourself.

After a while, though, a change would occur. The novelty would wear off, and what used to be fun would turn into yet another chore. Persistent that you are, you would endure for some time, pushing yourself to continue, until one day you would skip a class, then another one, and another one. Sooner or later, that fancy gear would end up gathering dust in the closet or in your basement, or worse yet, it would be standing in the middle of the room like that indoor bicycle, constantly reminding you of your character failure.

What happened there? Well, the good news is, it's nothing to do with the structure of your personality. What took place is that you just hit the plateau of disappointment of the proverbial learning curve (Fig. 5.14).

Remember the Self Determination Theory we mentioned in Chap. 3? It is a psychological framework that I like to refer to. According to this theory, we humans, are driven by the necessity to satisfy some basic psychological needs. These needs are as powerful as our physiological needs for air, water, food, and shelter. The two most relevant ones for the phenomenon that we are discussing are:

- autonomy,
- competence.

Simply put, we want to be active participants in our own lives. We have the urge to feel competent in what we are doing. Furthermore, we have an inherent need to see our skills improve and grow. It is an evolutionary strategy. An organism has a better chance of survival in an ever-changing environment if it is able to learn and acquire new skills. Our brains are hardwired to reward us with a little squirt of dopamine whenever we reach a new milestone on our learning path. The more often this happens, the better. It means that we are using our time productively by constantly improving. Ideally, we would be reaching new milestones in regular time intervals. In other words, our skills would be improving linearly with time. Unfortunately, this is not how the universe works. It is not called a learning curve without a reason. Our progress is non-linear. It is exactly the path that I describe, an S-shaped curve consisting of three distinct parts:

- slow beginning,
- the steep slope of progress,
- plateau of disappointment.

Initial determination and enthusiasm will help us plow through the initial slow start until we reach the slope of the steep progress; however, once the improvement milestones become too far apart, our brain will do a bit of mental calculation. It will try to estimate the ratio of effort versus the perceived achievement, and at some point, it will decide that our time will be better spent doing something else.

Bending the Curve

The same thing applies to games. Have you ever abandoned a game just because you got stuck on a particularly nasty level or a difficult boss fight? We play games because they are one of the most convenient ways of satisfying our psychological needs, see Chap. 3. However, if we are stuck at some point, it means that our skills in a particular game have reached that dreaded plateau.

So, as a game designer, what can you do to avoid this problem? What can you do to keep your players entertained by your game as long as possible? The key to this is to maintain the density of milestones that a player can reach in order to feel good about himself. This is the so-called Density of Goals (see my previous text). Basically, at any given point in time, a player needs to have a goal that is within his immediate reach. Obviously, this is very hard to do if goals are presented sequentially. Think of levels in games such as Angry Birds or Candy Crush. Instead, over the last few decades, another type of metagame has grown in prominence to address exactly this problem.

The letter RPG stands for Role Playing Game; this type of gameplay originated in a particular genre of story-driven games, which allowed players to develop their character during their journey through the game world. In order to allow for unique ways in which these characters can evolve, each character is represented by a series of properties known as stats, which can take various numerical values. For example, an RPG character can be described using the following set of properties:

- Speed—defining the maximal velocity at which it could move on the map or during the battle,

- Health—defining the maximal amount of damage it can sustain before being thrown out of the game,
- Stamina—denying the amount of time a character can run at the maximum speed,
- Precision—defining the probability of actually hitting a target when using a ranged weapon such as a bow,
- Luck—affecting the probability of finding valuable items when opening chests and similar things.

A character could also be equipped with various weapons and pieces of armor that could in some way affect his stats. Each of these items can have properties of its own. For example, a sword can be described using the following properties:

- Damage—defining the amount of damage it can inflict upon a target, i.e., the numeric value by which the health of the opponent is decreased each time it is hit with this weapon.
- Durability—defining the number of times that the weapon can be used before breaking apart.
- Value—the number of gold coins that the player might receive if he were to sell the weapon, etc.

These systems can become quite intricate, including a list of twenty or more properties. Typically, the stats, i.e., the numeric values assigned to these properties, can be in some way increased during the game (Fig. 5.15).

Ok, but how does this system affect players' goals? Instead of having a sequence of goals, i.e., a linear list of levels that the player needs to complete one by one, a player of an RPG game always has multiple choices. Each stat of each character becomes a mini-progression vector in its own right. The Player's goal is to upgrade any of the stats to the next level, i.e., to increase its value. The player can pick and choose which stats to focus on. Perhaps having a fast character is better suited for his playstyle, so he will focus on improving his Speed and Stamina. Or his character is lacking Luck, so he will focus on tasks aimed at improving this stat, etc.

The main advantage of this metagame structure is that each of these stats has its own improvement curve, consisting of milestones that the player can chase. This, in turn, reduces the effect of the player being stuck on a particular plateau of disappointment. If the progress path is blocked in one place, the player can always choose to pursue the improvement of some other stat (Fig. 5.16).

As I already said, this type of metagame evolved within a particular eponymous game genre. However, the game systems inspired by this approach have seeped into a multitude of other game types.

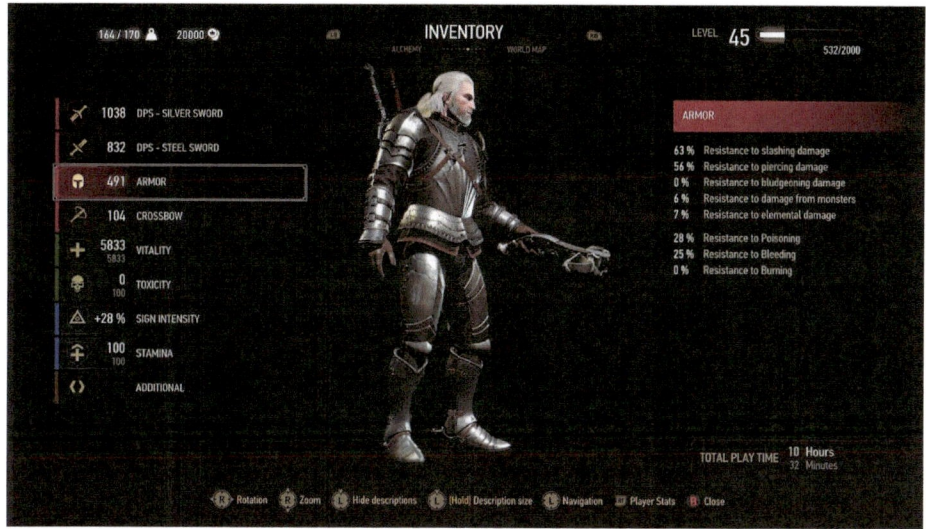

Fig. 5.15 A screen from Witcher III that shows the stats of a piece of armor

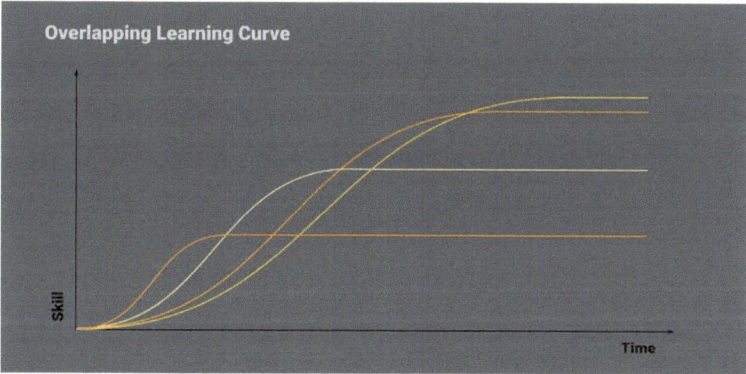

Fig. 5.16 Overlapping learning curves blur the impact of the plateau of disappointment

RPG Mechanics in Brawl Stars

Brawl Stars is a mobile game published by Supercell in 2018. It has so far amassed over $1 billion in revenue and has millions of daily active users. It offers a cheerful game world full of color. It is a real-time multiplayer brawler. Thematically and in terms of gameplay, it is a world removed from classical RPG games such as Diablo or Elder Scrolls. Still, its metagame applies a lot of RPG patterns.

Fig. 5.17 Character selection screen of Brawl Stars

First of all, just like any RPG, it revolves around characters. The characters known as Brawlers are the heart and soul of Brawl Stars. This is even indicated in the name. At the moment, there are a total of 52 characters in the game. Unlocking all of these is a powerful aspirational goal, but unlocking even a single new character is a major goal for every player. New characters are being added with regular upgrades. Each of these characters has four stats associated with it. Brawl Stars is not a "true RPG". It offers only a simplified upgrading system. Stats cannot be upgraded individually or directly, rather the character itself can be upgraded. Upgrading a character will result in an increase of at least one of its stats (Fig. 5.17).

Each character has 10 upgrade levels in total. Multiplied by the sheer number of characters, this alone creates over 500 distinct minor goals that the player can pursue! To upgrade a character, the player needs to have a specific amount of two resources:

- Power Points (pun intended),
- Coins.

Each character has its own pool of Power Points, while Coins are universal and can be used to upgrade any of the characters. Both of these resources can be obtained from two additional features of the metagame:

- Trophy Road,
- Brawl Pass.

These two features constitute two main progression vectors of the game.

Fig. 5.18 Brawl Stars Trophy Road

Trophy Road is a ladder of tiers. The player can climb this ladder by performing well in individual matches. The speed by which a player progresses along this vector is, therefore, fundamentally limited by the player's skill and the time that he is able to sink into playing the game. Whenever a player reaches a new reward tier, he receives a prize. These prizes have direct gameplay value that actually lets the player move along other metagame progression paths. They are either Power Points or Coins, which would allow the player to upgrade some of his characters (Fig. 5.18).

In addition to this, a player can occasionally unlock a new character adding yet another parallel line along which he can progress within the metagame.

Note that players can also lose trophies and move backward on the trophy road. This makes the Trophy Road a particularly challenging progression path.

An alternative is offered via the other main feature, i.e., the Brawl Pass. This feature is structured in a similar way as the trophy road. Again, it is a ladder of tiers that players can unlock by collecting points. The key difference between these two features lies in the way these points are collected. Brawl Pass requires Tokens, which can be obtained by completing specific quests.

Quests are tasks that require that a player accomplish a variety of small goals within various modes of the game, such as winning a specific event several times or dealing a certain amount of damage using a specific character, etc. The player can have up to 12 quests opened in parallel (Fig. 5.19).

All these features work in unison, reinforcing each other in a very intricate and tightly-knit metagame. The density of goals offered to players is staggering. Whenever a player starts a new play session of Brawl Stars, he is offered a smorgasbord of quests and goals he can pursue, bigger or smaller in scope, easier or harder to achieve. Chances of a player being stuck in this metagame structure are indeed very small, as there will always be at

Fig. 5.19 Brawl Pass in Brawl Stars

least some quest that the player will be able to complete or something to unlock during any given play session.

This is just a very brief summary of Brawl Stars metagame. In addition to these features, this game employs various other supportive features.

Sessionning

In order to be able to construct flexible play sessions, it is good to understand their common structure.

Let's assume that your player is familiar with your game and that he has been playing it already for some time. In this text, we'll willfully ignore the problems of FTUE. Let's also assume that your player is starting a new play session after taking a break. This break can be anything from a couple of minutes to even days or months.

As he has been playing your game previously, your player will have a certain level of familiarity with the game. Obviously, he will also have a reason for returning to your game. This can be entirely by a happy accident, or it can be something under your control. Perhaps it was a push notification. Perhaps the phone's OS notified him about a new update, or he spotted your new icon. Unfortunately, in most cases, you will not be able to know this reason.

No matter what the reason is, it is safe to assume that the player has forgotten what he was doing in his previous play session. At this point, the task for your game is not so much to remind the player about what he was doing last time when he was playing. The first order of business is to help the player to pick out a new goal! Returning player experience is about helping the player to choose a new goal, any goal!

Instead of thinking about what the player was doing previously, think about what you would like him to do now. What is the most important thing that you want to put into his attention? This is a new feature or new content that you just put into the game. It may be a new live vent that you are running. If you do not have anything specific to advertise to your player, then you can direct him simply to the core gameplay or let him choose his own metagame goal.

Meaningful Goals

Everyone wants to write their own success narrative. Each gameplay session needs to feel meaningful. In order to do so, the player needs to have one meaningful goal for each play session. To be meaningful, a goal needs to be:

- Attainable during the play session.
- Feel consequential in the larger context of the game.

Establishing a context for each goal is what metagame design is all about. It is a story bigger than the scope of this text. In order to be attainable, it means that the player needs to be able to reach the goal with the skills he has within the time at his disposal. Unfortunately, you probably know neither the player's precise skill level nor the time he can devote to your game. What helps, in this case, is the density of goals.

You can leave the figuring of the player's abilities and schedule to the person who knows the most about these things, to the player himself. Your game should provide the player with as many goals as possible. These goals should vary in difficulty and time requirements. The player will be able to choose at least one of them to chase during his play session. Ideally, he will be able to pick out as many as possible he could fit into the time that he has at his disposal.

These goals could be anything, from trivial, such as collecting resources in a crafting game to winning a race, beating a level, or unlocking a new reward tier in the season pass.

Structuring the game in this manner allows for flexible session lengths that can accommodate different play patterns and daily schedules. It also implies that your game needs to have a rich enough metagame that can support the required density of goals.

Cliffhangers

Eventually, the player would need to end the session. Perhaps his moment of free time will run out, or you, as a game designer, have a legitimate reason to make him leave. In any case, what the player needs is a safe moment to end the session. A moment when he feels that he has accomplished enough to be satisfied with the way he has spent his time. He also needs to feel like there are no other things in the game that need his attention. No one wants to leave a virtual city in danger of burning down while the mayor is away! In many cases, this means that there are no further actions available to the player.

This can be achieved with a variety of methods:

- *Explicit* means—such as energy mechanic, where UI indicates that the player has no more energy/fuel/action points left.
- *Implicit* means—in which the metagame is structured in such a way that there are no tasks left for the player to do.

If these conditions are met, the player will feel free to leave the game with peace of mind. However, as a game designer, you obviously want your player to eventually return! Your game will still need to provide some sort of a cliffhanger, some sort of return trigger, or a bit of unfinished business that will lure the player back in.

You are modeling the player's time both with the game or away from the game! When designing return triggers, you are actually modeling the player's time away from your game.

Here again, we encounter two general approaches. Some of the games rely on appointment mechanics that create return triggers for the player. Usual energy mechanics, found in many casual games, is a typical example. The player runs out of hearts or energy units, and he is prevented from playing for a certain amount of time until energy is regenerated by the system (Fig. 5.20).

The main advantage of these systems is their relative simplicity. On the other hand, these systems remove the sense of agency from the players, which is their main disadvantage. Such systems are generally perceived as restrictive and are not appreciated by the players.

The other approach is to allow the player to set his own return triggers. Typical examples of this are production loops in crafting games such as Hay Day or chest opening

Fig. 5.20 Out of Hearts in Candy Crush and Angry Birds 2

mechanics employed in Clash Royale. The key advantage of these systems is that they put control over the time in the player's hands. In a typical farming game, time management is an essential core gameplay puzzle. The player tries to optimize the usage of his time both during the play sessions and away from the game.

Mathematics of Game Design

6

Balancing a Game

There are many types of game designers. There are many areas of our discipline that you can specialize in. When asked what kind of a game designer I am, I usually tend to define myself as a systems designer. Making long-term retention features is my specialty. Perhaps my engineering background prepared me for this role, or maybe I just had such luck working on a lot of such features over the years.

One of the questions that I get asked very often is how you should approach balancing the economy in a complex free-to-play game. Especially how should you balance features that are intended to keep players engaged for a long period of time, weeks, months, or in some cases, even years?

This section describes the methodology I tend to use. It may not be the best, but it has served me well so far. To seasoned designers, this might seem trivial, but I decided to share it in the hope that someone might find it useful.

I will use a very simple example to illustrate my methodology. However, this approach forms a basis for balancing more complicated stuff. I will also use an example of an imaginary feature for an imaginary game, but I'll use a type of game that each game designer should be pretty familiar with.

Let's imagine that we are doing feature balancing for a real-time PvP game. For example, something that plays like Clash Royale. Players play 1 on 1 matches and gather some points. Let's call them Trophies as rewards. By collecting these trophies, players can travel down the so-called Trophy Path. You should be familiar with this type of feature as it is common in many games. Brawl Stars by Supercell is a good example of this (Fig. 6.1).

A Trophy path is a relatively simple feature. It is a ladder of reward tiers. To reach each subsequent tier in the ladder, the player needs to collect a certain amount of trophies. Our objective is to balance this feature. In other words, we need to come up with a set

© The Author(s), under exclusive license to Springer Nature Switzerland AG 2024   105
S. Stanković, *Game Design for Free-to-Play Live Service*, Synthesis Lectures on Image, Video, and Multimedia Processing, https://doi.org/10.1007/978-3-031-56156-6_6

Fig. 6.1 Battle Legion by Trap Light has one of the pretties implementations of the Trophy Road

of values, one for each point threshold, i.e., how many trophies in total a player needs to collect in order to unlock a particular reward tier.

The tool that we are going to use is spreadsheet software, Excel, or Google Sheets. It helps if you are familiar with those, but you can always use pen and paper.

General Guidelines

The first important thing that we need to understand is that our job is to actually model the player's time!

Time and not the points, the trophy count is the key to understanding the balancing of such features. A Trophy Road is a long-term retention feature, and the player is expected to be engaged with it for a very long time, as I already mentioned, weeks if not months or years. Balancing such a feature might seem daunting, but it is a rather straightforward exercise. Do not fear the large numbers!

In order to proceed, we are going to need to make some assumptions on how a player spends his time. If you are designing a new feature for an existing game, collected telemetry data might help you make you do this. If you are making a brand new game, benchmarking the competition titles can help. Some of these assumptions will prove to be wrong. This is expected and unavoidable. I will try to teach you how to err on the safe side and minimize the impact of errors.

Collecting Trophies

We begin our exercise at the source, at the source of Trophies, that is. The first thing we need to determine and fix is the way the player obtains the Trophies.

We are working with a 1 on 1, real-time PvP game. For each player, the outcome of a single match can be a victory, a tie, or a loss. Naturally, players will expect to get Trophies for every victory that they achieve. At least, this is simple. However, in free-to-play games, if we want to achieve long-term retention, it is good to reward the player's persistence and not only the player's skill.

The player should be able to advance along his progression vectors even if he is not particularly good at the game. The player should not get stuck. Having a skill should make the player progress faster. Therefore, I would also give a certain amount of trophies to players even after losing a match as a form of a consolation prize.

You can decide how you want to treat ties. You can treat them as victories and reward the full amount of trophies to each of the players as if they both have won, or you can give some fraction of that number, i.e. 1/2, as if they have split the prize.

In general, you are free to come up with exact numbers for both victory, loss, and tie. You can decide to give fixed numbers of trophies for each of these events or have some formula to calculate the appropriate amount. Elo score can be used for this, and it has the advantage that it is actually made for this type of game. We discuss the Elo rating scheme in detail in a separate section of this chapter.

Making Assumptions

As I said, what we are doing is modeling the player's time. What we actually want to do is to predict where along the Trophy Road our player will be at a certain point in the future, or if we put it the other way, how much time it will take for the player to reach a particular reward tier on the Trophy Road. The distance here is equivalent to the total accumulated Trophy Count. In order to do this, we need to make some assumptions.

Above all, we need to somehow estimate the speed by which the player will accumulate the trophies. Obviously, this depends on the number of matches that the player is willing to play in the given amount of time. You are likely going to have a huge spread of values as players differ significantly in their engagement levels. Some will play like crazy day and night, while others will play only a couple of matches per day. In order to deal with this spread, it is useful to make a couple of scenarios.

The first scenario I like to call the Insane Scenario. It usually assumes the theoretic maximum of points that anyone can accumulate by an insane amount of engagement. Let's assume that some imaginary, absolutely crazy player would spend 12 h each day playing our game. I usually chose 12 h instead of 24 h because it is reasonable to assume that even such a player has to eat and sleep during the day. 16 h is also a good choice sometimes.

This is not something that any actual living person would do, but it is useful to make this analysis as a sort of the upper limit, a theoretical cap on a player's speed. Let's assume that there are no limits on the number of matches beyond the actual length of the match. You can estimate the average length of the match based on your core game. Your game can also implement explicit match duration, e.g., each battle lasts for 2 min, etc. Dividing the number of hours played per day will give you the number of matches played per day.

You might want to implement some restrictions on this. Your game might have an energy mechanic or some other feature that limits the number of matches that a player can play during the day. If so, you need to take this into account also.

Assume that this insane player is also extremely good at playing the game and that he will win each match he ever plays. His win/lose ratio is 100%. Multiplying the number of matches per day with the number of trophies awarded for each victory will give you the number of points that this player collects per day.

You may also want to limit this number by some external means. Maybe you can implement a cap on the number of trophies that a player can win per day. Playing additional matches after the cap has been reached might not give any additional points.

You can now calculate the total number of points that the player will have on each day of engagement with the feature. This scenario gives us the upper limit of the player's velocity. It has its use, but it is not very realistic. It largely serves as a sanity check.

In order to make the actual balancing, we should make another more realistic scenario. I like to call it the Engaged Player scenario. Assume a more realistic session length and session count. Again if you are building a feature for an existing game, you are likely to have such data at hand. If you are working on a new title, you might try to obtain such data from some online source, such as SensorTower, AppAnnie/Data.ai, or GameRafinery.

Divide the total average playtime by the average length of the match to determine the number of matches that the player plays during each day. Assume some win/lose ratio. In a properly balanced PvP game win/lose ratio should, in the long run, be for each player $R = 50\%$. You can also assume any other ratio, for example, $R = 75\%$. Use this ratio to calculate the average amount of points the player is expected to take out of each match, whether he wins or loses, $R*V + (1-R)*L$, where V are points awarded for each victory, and L are points awarded as a consolation for a loss. Multiply this number by the number of matches played per day to obtain the average number of points that the player is expected to collect each day.

Similarly, as in the Insane Scenario, we can now calculate the total accumulated number of points that the player is expected to have at the end of each day!

Finally, it is good to make another scenario. Obviously, the polar opposite of an insane player with maximal engagement would be someone with minimal engagement. By definition, such a player would have 0 matches played during the day and 0 collected trophies. This is not really useful for us.

The third scenario would be someone who plays the game only casually. The casual player would play only a fraction of the matches that an engaged player would play. We can, therefore, just divide the number of played matches by some number and obtain what we need for this player. It is reasonable to assume that a Casual player would play 3 or 4 times fewer matches per day and that he would accumulate trophies 3 to 4 times slower.

It is good to take a look at the implied session length and session count of such players.

Distributing Tiers

The calculations we just did are basic mathematics, as dry as bones. The next part is much more fun. It is where the rubber hits the road. It is where psychology meets mathematics.

Sure, the player will be grinding trophies and moving along the Trophy Road, but we also want this activity to feel as pleasant and as rewarding as possible. This is a game, and it should provide entertainment. This is the only way to ensure long-term retention, which is the purpose of this entire feature.

Linearity is predictable and boring. Curves are much more fun (Luton, 2022). What we want to do is to allow players to unlock new tiers relatively quickly at the beginning of their journey. We want to let the unlocking of the tiers fall in a steady rhythm later on. By letting players unlock a lot of tears quickly at the start, we help them get emotionally engaged with the feature and the game more quickly. Later on, when the player is already committed and has already established his gameplay routine, we can make this rhythm slower and more predictable.

It would seem unusual if a player would unlock more than one reward tier after playing a single match. The number of points that a player earns for a single victory, or the average expected number of points that the player earns per match, provides us with the first building block.

What I usually do is let the player earn the first reward after playing one, maximally, two matches. I also make sure that the player actually wins those matches. The trophy threshold for the first reward tier I typically set is the number of trophies that the player gets for a victory.

I set the number of additional trophies that the player needs to earn for the next following tiers using the same value. I then gradually increase the trophy gaps between tiers until I reach a particular steady value, which I then repeat for the remainder of the trophy road.

When the rhythm settles, I like that my Engaged players are able to unlock a new reward tier about once per day. In other words, a day of relatively engaged gameplay

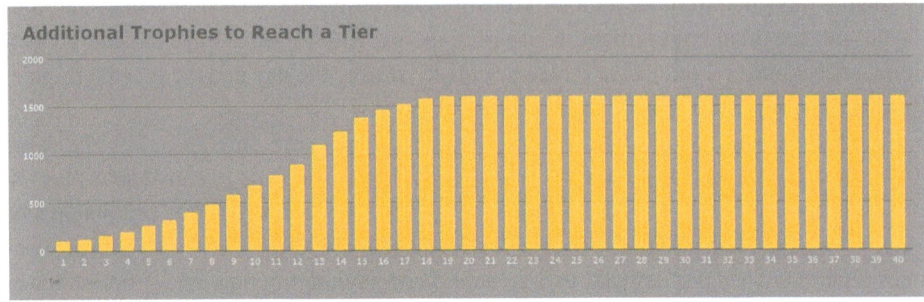

Fig. 6.2 Chart of additional trophy count needed to reach a particular tier

should earn you a new Trophy Road reward. Casual players would, of course, move through these tiers 2 or 3 times slower, meaning that a casual player would be unlocking one reward tier every 2 to 3 days, which is still not too bad.

Of course, depending on your preferences, the nature of your game, and the audience, you can make this rhythm faster or slower. You can use the velocity of the casual player as guidance. You can make it so that your Casual player unlocks one tier reward per day and that your Engaged player gets to unlock 2 or 3 tiers per day. There are two things to consider, though.

One is the content amount that you need in order to make these rewards feel meaningful. This is up to your production pipeline and its capacity.

Another thing to consider is directly related to learning how to err on the safe side. We made a lot of assumptions about players' behavior. By definition, these are unreliable. It might be the case that your assumptions are widely wrong and that you would need to rebalance your Trophy Road after it goes live. You are free to make things easier or harder for the player. However, making things easier always feels better than making them harder. Err, on the safe side. Make the Trophy Road thresholds as high as you dare. Later on, you can always make them lower! This is far more preferable than having to make them higher because most of your players are blasting through it at too fast a pace.

Finally, plotted on a graph, your point thresholds should form a nice S-shaped curve Fig. 6.2. This curve matches the shape of the proverbial learning curve of an average player (Loerich, 2016), i.e., the way a player's skill improves over time. In this way, it corresponds to the player's expectations. We are going to examine these two curves in detail in the next section of this chapter.

Distributing Rewards

Now that we have the Trophy thresholds determined and reward tiers distributed, it is time to finally do the fun part and place the actual rewards in those brackets. This is the icing on the cake. One thing to note here is that you must have rewards that seem attractive to your player. Second, you need to have diverse rewards. This can be harder than it seems.

Assume that you have a relatively simplistic metagame. One that revolves around character collection and nothing additional. Typically, this type of game is focused on character collection and deals with only a handful of resources, such as character shards and soft and hard currency.

You can use a bit of smoke and mirrors to create variety in your rewards. You can use character unlocking as a reward itself. You can give out bundles of several character shards for each individual character. In addition, you can give piles of soft currency and bits of hard currency. You can also take two or more of these resources and pack them in a loot chest. Further, you can create different sizes of loot chests. In this way, you can create an illusion of the variety of rewards you place along your trophy road Fig. 6.3. This is just a guideline and not a rule. Some games get away by simply placing gacha boxes on such features.

In general, I like to think about rewards by dividing them into two categories:

- juicy *prime rewards*,
- *filler rewards*.

Prime rewards are the stuff that is attention-grabbing. These are the things that any player would want to have. In our example, unlocking a new character would be one typical prime reward. These are rewards whose value can be easily perceived at first glance. Bits of hard currency are also a good example of such rewards. Big gacha boxes can also act as such. These rewards will act as short-term aspirational goals. On the other hand, the

Fig. 6.3 Variety of rewards in the trophy road of Zooba by Wildlife Studios

filler rewards are not simply stocking stuffers. They are a foil for the prime rewards. They are there to serve as a contrast, underlying the value of the Prime rewards.

Calculate how many primer rewards you have or can afford to produce. Place them in a steady, simple pattern onto the Trophy Road. Every 4th or 5th tier can be a good rule of thumb. Ideally, an engaged player would be at the beginning unlocking one such reward each day and once a week later on. In this way, these rewards can really serve as midterm aspirational goals.

I like to place the first prime reward into the second or the third tier of the trophy road. I place some filler rewards as the first one that the player wins. This is deliberate. I want to reward a player for the first victory. The act of collecting any reward is in itself rewarding. The first juicy reward should require some persistence and repeated effort. Not too much of it, as I want it to be within reach for even the Casual players.

Since a trophy road typically has a lot of tiers, you will likely implement it using some scrollable UI. When designing it, make it so that the player will always be able to see at least one prime reward on the screen. In other words, the relative distance between two tiers offering prime rewards should not be bigger than one screen Fig. 6.4.

Once you have the prime rewards placed in this pattern, fill in the remaining tiers with filler rewards. Scatter them around to make the trophy road feel rich and diverse.

An important thing to consider when rewarding bundles of soft and hard currency is the amounts that you are giving! As always, in free-to-play design, you do not want to inflate the currencies by giving out too large amounts. Also, any reward system is, by definition, an extrinsic reward. Players should play your game because the game itself is fun. You should not bribe them to play it.

To determine the right amount of currencies to give, it is good to again think of time. A player's progress through the Trophy Road is effectively a measure of his effort, i.e., it

Fig. 6.4 Prime rewards on display in Trophy Road in Brawl Stars

reflects the amount of playtime. Use the velocity estimates to decide how much of each currency you are willing to give to a player for a specific amount of playtime.

In this way, you are establishing a correspondence between each of the currencies and the playtime. If you are making a feature for an existing game, you most likely already have these ratios already established. Furthermore, you can use your telemetry data to derive an understanding of player spending patterns. If you are building a new game, take this new source into account when designing currency sinks. A player should be expected to spend all the currency he gets in the same amount of time.

Finally, once you arrive at a set of values and the configuration that you are reasonably happy with, learn from your system. Playtest it and see how it feels. Try out different tuning of the parameters. Keep in mind, though, that making small incremental changes might work well for finetuning. To truly learn from your own system, you should try it out to see how it feels by making more drastic changes. Multiply everything by two or divide it by the same number. Try to see how things feel if you repeat this by a factor of 10.

The methodology I have described focuses on one type of feature in one type of game. It is, however, applicable more or less directly to other similar types of features, such as premium passes or limited-time event tracks. The general approach, which takes into account gameplay time and velocity of point grinding, is applicable to a much wider variety of features.

Rebalancing

As I have already pointed out, any game balance you make before actually releasing the game or the feature is going to be based on a lot of assumptions about players' behavior. These assumptions might turn out to be incorrect. Luckily, operating a game as a live service offers both a chance to learn from the real telemetry data and to fix errors.

If it turns out that your initial assumptions were not correct, you might find yourself in a situation where you want to rebalance and change the values of some variables. This is actually quite a common situation and is routinely done. However, keep in mind that the direction of the change matters. In general, if something is required of a player, you can always decrease the values. If something is given to a player, you can always increase the values without the risk of a backlash. For example, you can always make prices cheaper, and very few players might complain. Conversely, you can always require point requirements smaller to make it easier for the player. For such variables, always start with as big values as possible and reduce accordingly if needed to rebalance. On the other hand, you can't really reduce the amount of things you give out as rewards unless you compensate the players with something else. Start with as low values as possible and increase them if you deem that this is needed.

S-Curve and P-Curve

One of the most typical tasks in game design is balancing some sort of a progression vector. For example, you might need to come up with a set of values, a number of points that a player needs to collect to advance to the next level. Another example might be the number of cards needed to upgrade a character. It could also be the number of tokens that a player needs to collect in order to unlock another tier in the premium pass or a limited-time reward track.

What we are talking about here are two sets of values that can be derived one from the other:

- Point thresholds, i.e., the total cumulative number of points that the player needs to unlock a particular level.
- Additional points that the player needs to collect in order to unlock the next level, i.e., gaps between levels.

Assume that the player starts with 0 points at Level 1. Let S_1 be the number of points that the player needs to unlock Level 2, let S_2 be the number ADDITIONAL of points that the player needs to unlock Level 3, etc., etc.

Assume, also, that the player starts with 0 points at Level 1. Let P_1 be the number of points that the player needs to unlock Level 2, let P_2 be the TOTAL number of points that the player needs to unlock Level 3 etc., etc.

It follows that $P_1 = S_1$, $P_2 = S_1 + S_2$, $P_n = S_1 + S_2 + \cdots + S_{n-1}$, and $P_2 = S_1 + S_2$. Likewise, it follows that $S_n = P_n - P_{n-1}$, $P_n = S_1 + S_2 + \cdots + S_{n-1}$.

When you are doing this sort of balancing task, you can always take a lazy approach and make each subsequent level require exactly the same number of additional points to unlock.

For example, a player would start at level 1, level 2 would unlock once the player collects 100 points, level 3 at 200, level 4... 300, etc. It is obvious that the gap between point-level requirements remains constant and is 100 points. The cumulative number of points that the player needs to unlock a particular level is a linear function.

This is very simple maths that is easy to follow and code. However, it is also very boring. Its intuitive nature is what makes it so. Even if we, as players, don't pay attention to numbers, our brains do. The human brain can easily subconsciously perform linear extrapolation. This predictability is what makes linear progressions feel monotonous.

Furthermore, as the game progresses, it is very likely that the player's ability to collect points will evolve. In skill-based games, the player is likely to become better at playing the game. In addition to this, the gameplay itself should and will evolve. This makes finding a good balance all the more difficult. A gap of 100 points might be daunting at the start of the game, while it might become trivial by the time the player reaches the end game. This is exactly the opposite of what you, as a game designer, want to achieve. The game should feel relatively easy at the start when the player has not yet mastered it and remain challenging forevermore. In order to achieve this, you should resort to a just a bit more complex type of math formula, i.e., non-linear functions (Aponte et al., 2009).

As time progresses, the player should need more and more points to collect in order to unlock the next level. The cumulative number of points should grow relatively slowly at first and pick up speed as the game unfolds.

This can be achieved with the use of an *exponential function*, a.k.a. The *power curve* of the *P-curve*. We can apply this sort of mathematical function to create. The picture below gives an example of one such curve. You can notice that the slope of the curve starts relatively gently and becomes steeper and steeper between levels 7 and 11, as shown in Fig. 6.5.

The power curve gives you, as a designer, a good overview of the progression in general. However, things are a bit different from the player's perspective. The players will most likely never see this curve as a whole. The player will most likely get to see the additional number of points he needs to reach in order to unlock the next milestone.

The player's behavior will be influenced greatly by the values of these gaps between thresholds. Having too big gaps at the start of the game would present way too big of a challenge for a novice player. Even if we are designing a live event aimed at seasoned players, most likely, we would like to make the start of the event easier. The point gaps between levels should, thus, grow as the player progresses through the levels. On the other hand, we do not want to let this increase continue to infinity, as we do not want to make the player have to deal with absurd values.

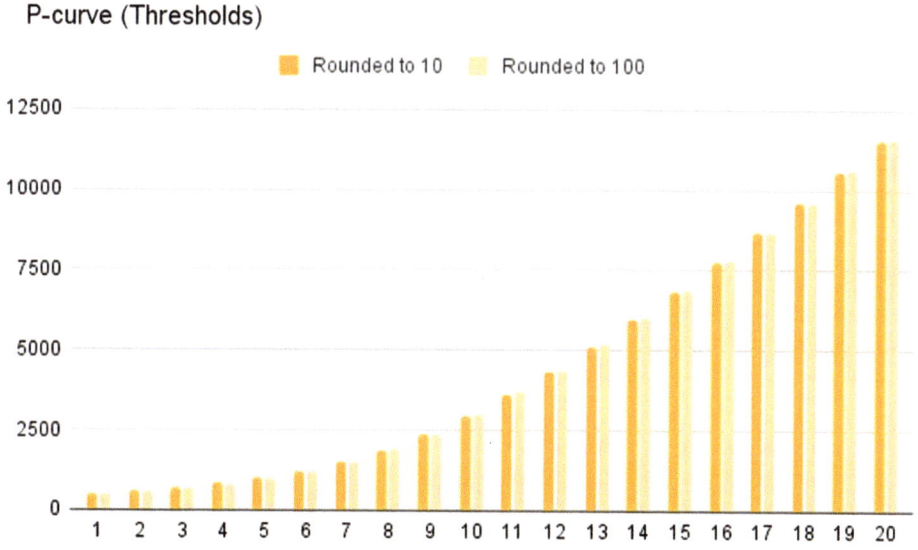

Fig. 6.5 An example of P-curve values

In order to achieve this, game designers usually apply what is known as the Sigmoid function or S-curve.

$$S(x) = \left(\frac{1}{1 + e^{-kx}} \right)^a$$

where k and a are parameters dictating the shape of the curve. Figure 6.6 shows an example of one such S-curve. Obvious $S(x) \in \{0, 1\}$. Such values are typically not directly usable as level thresholds. In order to get some more usable values, I applied the scaling formula $S'(x) = l + m * S(x)$, where $l = 100$ there is an arbitrarily selected starting value and $m = 1000$ a scaling factor, also selected arbitrarily.

As you can notice, the values are nearly constant at the very start. They begin to grow faster at around level 5 and grow at the fastest rate until level 10. The slope becomes shallower again. In the end, values become constant from level 18 onwards, as shown in Fig. 6.6.

In many ways, the S-curve approximates the learning curve that usually follows the acquisition of every new skill by the players, which we discussed in Chap. 3. The idea here is to maintain the constant rate at which the player reaches milestones within the game, maintaining the so-called density of goals.

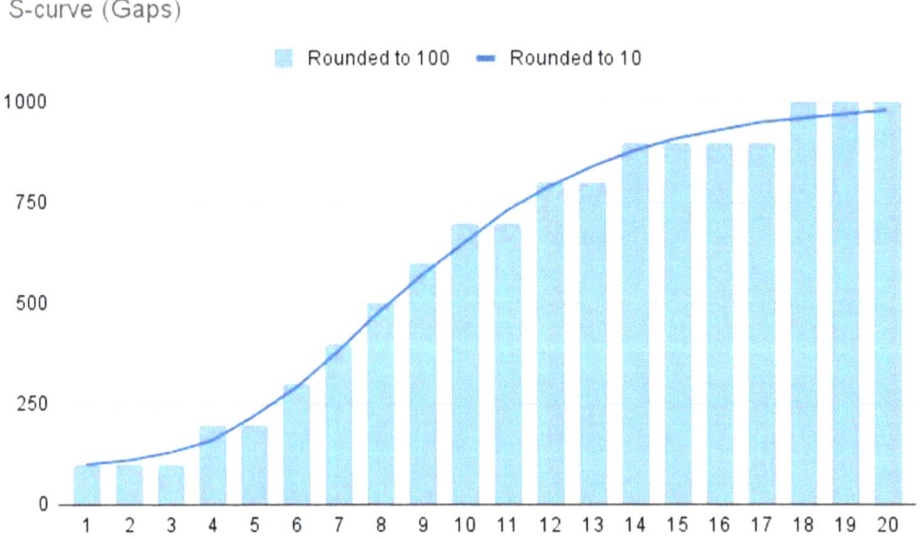

Fig. 6.6 An example of S-curve values

Scale and Absolute Values

The previous discussion describes essentially just the shapes of the curves, leaving a set of parameters k, a, l, m with which you can play in order to generate final values. However, it is not just the shape of the curve that affects the player's perception. The actual values also matter.

Believe it or not, numbers also have a certain aesthetic. If you are creating a series of values that will be exposed to the player, you might want to make them easy to understand. Round the values, and you get to the nearest increment of 5, 10, or 100. This is more user-friendly and easier to comprehend and remember. This is exactly what I did with the examples from the previous section. The raw values produced by the formula and rounded values to the nearest increment of 10 and 100 are shown in Table 6.1.

Table 6.1 Values obtained for $S'(x)$

Samples	Raw values	Projected on Min/Max	Rounded to 10	Rounded to 100
1	0.003912602816	103.52	100	100
2	0.01258965914	111.33	110	100
3	0.03299039741	129.69	130	100
4	0.0718751514	164.69	160	200
5	0.1334360717	220.09	220	200
6	0.2165841539	294.93	290	300
7	0.3149960337	383.50	380	400
8	0.419638555	477.67	480	500
9	0.5218184139	569.64	570	600
10	0.6151596104	653.64	650	700
11	0.6961482846	726.53	730	700
12	0.7637071078	787.34	790	800
13	0.818405231	836.56	840	800
14	0.8617047825	875.53	880	900
15	0.8954065227	905.87	910	900
16	0.9213081899	929.18	930	900
17	0.9410280172	946.93	950	900
18	0.9559361872	960.34	960	1000
19	0.9671479447	970.43	970	1000
20	0.9755470861	977.99	980	1000

Please do not overlook the scaling parameters l and m. The granularity of the values matters. For whatever reason, you might encounter a situation when you would like to insert a series of values between existing thresholds. For example, you are designing a new limited-time event and would like to reuse the basic set of values generated for a previous one. However, you want to introduce more reward tiers without a need to change every value.

If you set values to low numbers, it is hard to insert new values in between or to divide them to lower them. Obviously, this is impossible if your values are too small, such as 1, 2, 3, ... If they are 10, 20, 30 ... you have some freedom to create a new set of variables that are more granular: 10, 15, 20, 25, 30 ... But if you try to do this again, things become weird: 10, 12.5, 15, 17.5, 20, 22.5, 25, 27.5, 30. Things look much better if you multiply them, 100, 125, 150, 175, 200, ... don't look weird. The bigger the original values, the more space to insert new steps you have.

Furthermore, the scale also matters. Smaller values such as 1, 2, ... 100. Implay that something is rare. Rarity implies scarcity, which in turn implies value. Hard currency values are usually in hundreds, not thousands or tens of thousands. Bigger values imply abundance. In old arcade games, you don't get one point for shooting an alien. You get 1000 points. Final scores are in millions, although the player has not short millions of enemies.

These observations apply to any values you might be working on regardless of the formula you used to generate them, i.e., the curve that they follow.

Beyon One Variable

The previous two sections discussed the basics of the very foundations of the game system design. Pointed out that you should begin by considering time as the key parameter of the game balance. We discussed how to select an appropriate formula to obtain the desired values and how to scale them. However, you will notice that we have focused solely on generating values for a single independent variable. Granted, a system consisting of only one variable is not much of a system.

The true challenges begin when more elements come into play. You might recall that internet meme saying that there are 915,103,765 ways to combine six 2×4 LEGO bricks (Nissen, 2004). But where do those mind-boggling numbers come from?

Games can be seen as a system of elements. Consider a system consisting of three elements A, B, and C. If these elements are represented as nodes in a graph, there are exactly three different connections between two of these nodes: (A, B), (B, C), and (C, A). There are also 8 possible ways how a graph can be constructed out of these nodes, as shown in Fig. 6.7.

If we generalize these numbers, we get that for a set of n elements, there are $n * (n-1)/2$ possible connections between two elements and $2^{n*(n-1)/2}$ possible graphs.

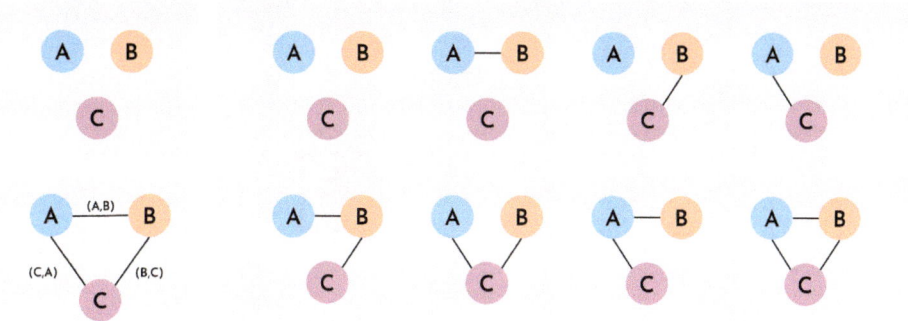

Fig. 6.7 All possible graphs with three nodes

Table 6.2 Number of possible graphs with n elements

No. of elements	No. of connections	No. of graphs
1	0	1
2	1	2
3	3	8
4	6	64
5	10	1024
6	15	32,768
7	21	2,097,152
8	28	268,435,456
9	36	68,719,476,736
10	45	35,184,372,088,832

That is to say that as the number of elements grows linearly, the number of possible ways to combine them grows exponentially! Table 6.2 shows how these values look as the number of elements grows from 1 to 10.

Keep this in mind whenever you are adding something to your game system!

Optimizing Two Variables

Did you ever notice how people play Angry Birds games? In the classic Angry Birds game, the player tries to solve a puzzle. The puzzle consists of figuring out which spot on the piggy tower should be hit by a bird to cause the whole structure to collapse. The player controls the slingshot. Essentially, what a player needs to do is to figure out the angle and the speed by which the bird should be launched in order to hit the appropriate target.

Fig. 6.8 The mathematics of classic Angry Birds

This is a two-variable optimization problem. Two-variable optimization is a difficult task. Human brains are not especially adept at it. Most players start by fixing one of these two variables and try to guess the other one. They usually start by always shooting at the full power and try to find the optimal angle. They start to optimize for the speed, i.e., the distance to which the bird would fly, only after they decide on one angle. Figure 6.8 shows an example of this.

This doesn't mean that human brains are, in some sense, flawed. Optimizing more than one variable is a mathematically difficult task. Most multivariable optimization algorithms work in a similar fashion. They temporarily lock all but one variable to some initial value and try to optimize it to some criterion.

Managing Complexity

You can apply the strategy of Angry Birds players when constructing the system with more variables. Select one system variable as the principal one. Most likely, this is going to be a variable corresponding to one of the most prominent progression vectors, see Chap. 5. The player's experience (XP) level is typically a good choice. Select an appropriate curve and create a starting set of values as described in the sections above.

Use this structure as a backbone of your system design. Balance all other variables in reference to this one. To manage complexity, try to create self-contained loops. In other words, try to limit the number of ways various parts of the system interact with each other. Such isolated loops are typically easier to balance.

SimCity BuildIt was, like so many other games, plagued by the inflation of Simoleons, its soft currency. In this game, a player is able to gain Simoleons from taxes collected

Fig. 6.9 Regions of SimCity BuildIt

from the player's main city. The amount received is calculated based on the size of the population of the city. Tax money is generated over time, requiring no other action on the player's part except logging into the game. In addition, we wanted to expand the playable area of the game and diversify the type of cities that players could build. The proposed design concept called for the addition of five new regions, i.e., satellite areas to the main city. This would increase the total possible population of each city several times, further exasperating our Simoleon inflation. Figure 6.9 shows regions of SimCity BuildIt.

To avoid this problem, I introduced a separate regional soft currency for each additional region. These currencies behave much the same as Simoleons, but all their sinks and sources are contained only within their native regions, i.e., they constitute a self-contained loop largely isolated from the rest of the game.

However, the isolation couldn't be complete. The player still needed to be able to purchase certain things, in this case, items and buildings, that originated from the main city. Creating a totally new and different series of prices for each item and building in each of these five new currencies would have created virtual financial chaos within the game, potentially leading to unforeseen consequences. In order to avoid this, I used two particular tricks. First of all, I decided that all five regional currencies would have exactly the same relative value. Sure, they would be represented differently and would function each in its own special region. However, their values would match. The second trick was to derive their relative value based on the value of the Simoleons, i.e., the main currency. The Simoleon prices of buildings and items were long established within the game. Players are used to them, and the pricing worked reasonably well. Still, copying one-to-one felt somehow wrong. Instead, I divided each price by a factor of 10, creating a new set of values.

This is the trick you can use. Introduce new variables by reusing existing ones. You can start by taking the existing well-balanced variable and multiplying its values by some factor. Choosing a factor can produce various results. If you multiply original values by some round number 10 or 100, etc., it makes it easier for players to understand the ratio. Depending on the situation, this might be what you want to achieve.

You may even want to keep the original values and restrict the change only to visual reskinning of the variable. Points in limited-time events are often done in this way. For example, players might be collecting shamrock points for St. Patrick's event and hearts for St. Valentine's. One shamrock equals one heart.

On the other hand, if you deliberately want to obfuscate things, you might want to multiply the original values with something that is hard to perceive at first glance. Prime numbers like 7 or 13 are great for this. For example, let the original values be 5, 10, 15, 20, 25, etc. It is obvious that 50, 100, 150, 200, 250, etc., are the original values multiplied by 10. On the other hand, it is hard to tell the relationship between 65, 130, 175, 260, 325, etc., and the original values unless you actually sit down and do the math.

Two Resources

Many games employ systems that rely on the interplay of two resources instead of one. The addition of another resource allows for a more flexible system and more variation in the gameplay.

An example of such a system is the troop upgrading logic in Clash Royale. Two resources are needed to upgrade each of the troops in the clash royale: Gold, i.e., soft currency, and the appropriate cards. In this game, Gold is a universal resource that is required for upgrading all troops. On the other hand, the cards are troop-specific, as shown in Fig. 6.10.

A new player starts the game with a relatively big initial amount of Gold, while the card supply is tightly controlled. This means that in the early game, cards are a scarce resource. Players are trained to perceive cards as valuable and Gold as an expendable resource. However, as the game progresses, the demand for gold grows drastically, driven by two factors. Above all, each subsequent upgrade of each troop type demands more Gold. Table 6.3 shows the amounts of cards and gold needed to upgrade a common troop type. These values are also shown in Fig. 6.11.

This is multiplied by the sheer number of troup types that get unlocked later in the game. By the mid-game, the relative value of the two resources flips, and scarcity of Gold starts to hit. A typical elder player has an inventory of troops that are ready for upgrading but lacking the required Gold. The player faces a choice, whether to upgrade the less used troup of the lower level with a cheaper upgrade cost or to try to save up the Gold and use it to upgrade a more expensive higher level troop.

Clash of Clans offers another example of the application of the two resources, Gold and Elixir. Both of these resources behave much the same way. They are both produced over

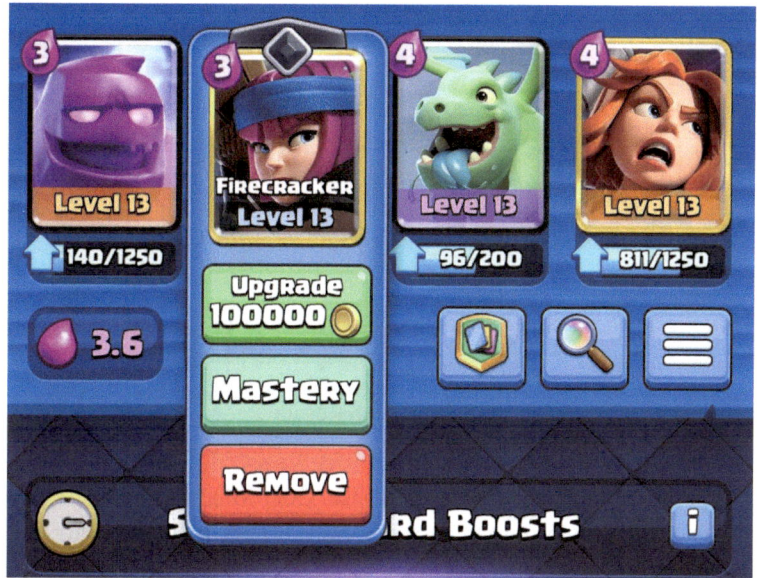

Fig. 6.10 The upgrade cost of a card in Clash Royale

Table 6.3 Upgrade cost of common troops in Clash Royale

Level	Cards	Gold
1	1	0
2	2	5
3	4	20
4	10	50
5	20	150
6	50	400
7	100	1,000
8	200	2,000
9	400	4,000
10	800	8,000
11	1,000	15,000
12	1,500	35,000
13	3,000	75,000
14	5,000	100,000

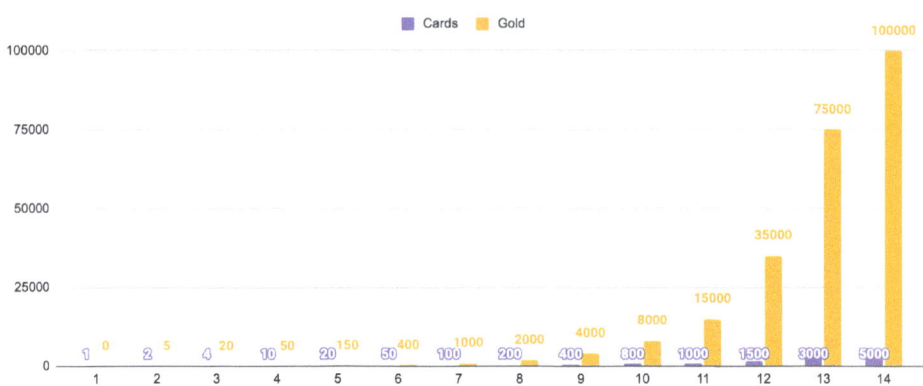

Fig. 6.11 Upgrade cost of common troops in Clash Royale

time by the special buildings in the player's base, Gold Mines, and Elixir Pumps, respectively. The player's ability to store these resources is limited by the capacity of the storage facilities, Gold Storage, and Elixir Collectors. All of these buildings can be upgraded in order to increase their production and capacity. The catch is that Gold Mines and Gold Storage require Elixir for their upgrades, while Elixir Pumps and Elixir Collectors require Gold for the same purpose, as seen in Fig. 6.12.

This forces the player to zigzag through the metagame, upgrading in turn first the production capacity of Gold, before in turn being able to upgrade the production capacity of Elixir and vice versa.

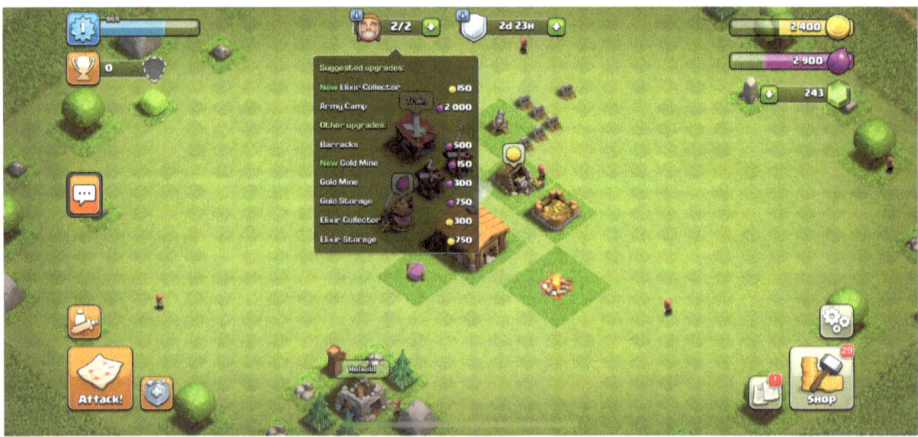

Fig. 6.12 Screenshot from Clash of Clans tutorial showing suggested building upgrades

Elo Rating System

In this section, we are going to turn our attention to something known as the Elo rating system. This is probably the most mathematically intensive part of this book. However, this is not a mathematical book, so I am going to keep the language deliberately sloppy and hopefully somewhat lighthearted.

In game design, the term matchmaking denotes the process of selecting two or more opponents to take part in a single match or a round of competition. In PvP games, any player who joins the game and wants to take part in a match is placed into a pool of eligible contestants. The system tries to find a suitable opponent for each eligible player according to some criterion. The objective is to provide an entertaining experience to both players. This means finding a suitable opponent with just enough skill to provide just the right amount of challenge. Matching a noob with a pro would result only in frustration for the noob and boredom for the pro. Skill is quality. It is often intangible. Still, computers work with numbers. We need to associate some quantitative numeric value with the skill of each player and use this as a matchmaking criterion.

The Elo rating system is designed to do exactly this, i.e., to associate a numeric value to each player that somehow reflects the level of the player's skill. What is more important, it provides a way to constantly update this value to reflect the improvements in player's performance. Despite the common misconception, Elo is not an abbreviation. It is actually the last name of Arpad Elo, a Hungarian-American physicist, mathematician, and chess player who created this system (Elo, 1967).

The system he designed was meant originally specifically to be applied to chess. Chess is, of course, a highly skill-based game. It is also what is known as a PvP 1:1 zero-sum game, meaning that each chess match involves exactly two players, presumably human and that there can be only one winner of each match. This makes the Elo system applicable to any other similar games. There are plenty of texts on the Elo rating system around. Some of those texts are more theoretical than others. There are also many modifications and extensions of this system for various particular other types of games.

What follows is my personal interpretation of the mathematics behind this system. I tried to provide practical step-by-step instructions in building this system, providing just enough theoretic considerations needed for understanding its workings. I also tried to reflect on actual issues and problems that I have encountered when trying to apply this system to games and game features I was working on. As you will notice, some of the presented values are somewhat arbitrary. They are mostly chosen empirically, and I have picked them up from the existing implementations of the system.

Basic Implementation

Imagine that two players, we shall denote them as player A and player B, are playing a PvP match against each other. These two players will have their corresponding Elo ratings, R_A, and R_B respectively. Before the match, the difference between the Elo ratings of these two players should be small enough for them to be selected as good competitors. In other words $|R_A - R_B| \leq M$, where M is the *matchmaking distance* of the players?

From the perspective of player A, this will result either in a *victory*, *loss*, or a *draw*.

The Elo rating of each player should reflect the player's relative skill, meaning it should reflect the player's past performance in matches against other players. Therefore, the Elo scores of both players need to be updated after the match in order to reflect its outcome.

Obviously, not all victories or losses have equal importance. Mercilessly beating a much weaker opponent doesn't count as quite the same kind of achievement as carving a victory against a superior adversary. On the flip side, there is no shame in losing a battle against a much stronger opponent, which can't be said for losing against someone with a much smaller rating. These should be reflected in the updated ratings of both players. Elo formula is designed to do precisely that.

After each match, the Elo rating of player A is updated according to the following formula $R'_A = R_A + K(S_A - E_A)$, which R'_A represents the new, updated Elo rating of the player after the match, R_A represents the Elo rating that the player had before the match, E_A denotes the expected outcome of the match and S_A is the actual outcome of the match. Variable K denotes a *scaling factor* that determines how much influence each particular match can have on the overall Elo rating of the player. There are several ways to derive this value. In practice, the factor K has a different value in different levels of competition, for example, in different leagues of the competition. A good starting value is $K = 32$ with which you can start to experiment.

As we mentioned before, each match can have only three possible outcomes from the perspective of player A. This is reflected in the values that the variable S_A takes in the case of victory $S_A = 1$, in the case of loss, $S_A = 0$ and in the case of a tie $S_A = 0.5$.

The true magic of the Elo formula comes from the second part, i.e., from the way the value E_A is calculated. Since $0 \leq S_A \leq 1$ it follows that, it would be nice that likewise $0 \leq E_A \leq 1$. Elo formula maps the expectation of the outcome of the match onto the interval $(0, 1)$.

$$E_A = \frac{1}{1 + 10^{(R_A - R_B)/c}},$$

where again R_A, and R_B represent the original Elo ratings of which players A and B had before the match. The factor c is another of the magic numbers in this calculation and is usually taken to have the value of $c = 400$. You can select any other value you want, but if you are uncertain, 400 is a good start.

You can safely ignore the following, but formally, the formula E_A is the logistic curve with base 10. It can be written as $E_A = \frac{Q_A}{Q_A + Q_B}$, where $Q_A = 10^{R_A/c}$ and $Q_A = 10^{R_B/c}$.

Now we have all the ingredients for the barebones implementation of the Elo rating system:

- The current Elo rating of a player R_A,
- The expected outcome of a match $E_A = \frac{1}{1 + 10^{(R_A - R_B)/c}}$ $0 \le E_A \le 1$,
- The actual outcome of the match $S_A = 1$, in the case of a win, $S_A = 0$ and in the case of a loss, and $S_A = 0.5$ in the case of a draw,
- The formula how to update the Elo score of the player after the match $R'_A = R_A + K(S_A - E_A)$,
- Two free scaling parameters K and c, with their typical values $K = 32$ and $c = 400$, and
- The matchmaking distance between players M.

Notice that since both $0 \le S_A \le 1$ and $0 \le E_A \le 1$ the biggest value by which Elo rating of the player can be updated equals K.

The formula by which the Elo rating of the other player, player B, is updated is, of course, exactly the same, only substituting R_B for R_A when appropriate.

One last thing remains that should be taken into account. This is the initial Elo rating of a new player who has just started playing the game and has not yet won any matches. Let's denote it as R_{A0}. A good value is $R_{A0} = 1500$.

Actually, any value can be chosen for this as long as it is commensurate to the maximal rate of change, i.e. K. Keep in mind that, in theory, a new player can start and, furthermore, can continue losing matches immediately after beginning to play. Fu Therefore, the new value of the Elo rating can be $R_A < R_{A0}$. This is something that we should try to prevent with other aspects of our game design, but our mathematical system needs to be able to handle this case. Finally, we do not want a player to have a negative Elo rating. The initial value R_{A0} needs to be significantly larger than the scaling factor K to allow for the player to play and lose enough matches before eventually winning one so that his Elo rating would finally start to rise up.

Just to be on the safe side and to limit negative user experience, it is good to introduce a minimal Elo rating value, R_{min} below which the Elo rating of an individual player cannot drop. In other words, if $R_A = R_{min}$ the player's rating will be updated only upwards!

Keeping Score

Are you still with me? Good! We are going to complicate things even more. The original Elo formula was designed for chess games. Chess is a game where players really do not keep score. Each match really can end only with three possible outcomes: win, loss, or a tie. Chances are that this is not enough for the game you are designing. Chances are that in your game, players will actually score during the match. Consider, for example a game of football, the European version known as Soccer in the US. A victory in which one team scores 5 goals against a team that scored 0 is surely a greater achievement than a victory of 1:0. The same applies to our game. What if we want to take this into account when updating the Elo rating of players after each match? There are at least two ways of doing this. One way is elegant. However, the other one is less elegant but offers more control.

The elegant way involves modifying the way that the numeric value for the actual outcome of the match S_A is calculated. Previously, we have used a very simplistic of calculating S_A by assigning fixed values for victory, loss, and draw. We can make this more sophisticated and take into account the scores achieved by both players during the match. Remember, $0 \leq S_A \leq 1$ it should remain. Let P_A be the number of points scored by player A and P_B be the number of points scored by player B during the match. We can calculate the desired numeric value as $S_A = \frac{P_A}{P_A + P_B}$, i.e., the number of points scored by player A divided by the total number of points scored by both players during the match. Obviously, in the extreme case, if player A has scored 0 points, $S_A = \frac{0}{P_B} = 0$ if player A has won and player B scored 0 points, $S_A = \frac{P_A}{P_A} = 1$ and if the match ended in a draw, $S_A = \frac{P_A}{2P_A} = 0.5$. Be careful of division by 0. In the case that no player scored any points $S_A = \frac{0}{0}$ should still produce $S_0 = 0.5$. In any other case S_A, it should result in a more granular value. For example, a 3:2 victory produces a value of $S_A = \frac{3}{3+2} = \frac{3}{5} = 0.6$ a value a bit bigger than 0.5 but still not quite as big as 1. On the other hand, a near loss of 2:3 will result in $S_A = 0.4$ a value a bit smaller than 0.5. This new way of calculating S_A can be neatly plugged into our original formula

$$R'_A = R_A + K(S_A - E_A).$$

However, this approach has some disadvantages. Above all, the results can be somewhat unpredictable. As game designers, we might want to retain a bit more control over the way that the Elo rating gets updated. Instead of plugging in the player score directly into the original formula, we might want to extend it in the following way:

$$R'_A = R_A + K(S_A - E_A) + L\frac{P_A}{P_A + P_B},$$

where again $\frac{P_A}{P_A + P_B}$ is the contribution of points scored by both players during the match and L is a new *scaling factor* that we are free to control independently from the rest of the calculation.

Rewarding Victory

There is one more final modification that I would make to the traditional Elo formula, and this one is not for the faint of heart. The Elo rating system is a piece of delicate math, well balanced and adjusted for the purpose it was designed for. However, this is also its biggest disadvantage when it comes to its use for game design.

The rating of an individual player actually reflects the player's skill level remarkably well. It has a tendency to grow at about the same rate the player's actual skill grows. Unfortunately, as we know, players' skills tend to follow a familiar S-shaped learning curve. This means that eventually, both the player's skill level and his Elo rating will reach a plateau, i.e., they will reach a point beyond which they will not grow significantly. This is all well and fine if we are talking about something like chess. However, we are not dealing with the grandmaster level of competitive chess here. We are most likely designing a game that is supposed to be, above all, entertaining for each individual player.

Quite often in games, we tend to use the player's level as an unlocking criterion for all sorts of things, from new game features to new characters, unit types, cards, arenas, etc. If we choose to use Elo rating as a proxy for a player's level, we are running a risk of having the majority of players getting permanently stuck at some mediocre level with very little hope of progressing beyond that point. Their true skill would simply reach the plateau.

In order to avoid this, we need to modify the basic Elo formula. I am going to take that sophisticated piece of mathematics and hammer it a bit with a sledgehammer to make it fit my purpose.

Let's analyze a bit the problem that we are dealing with. Consider a basic 1:1 PvP scenario. Each match has exactly two participants. Each match can have exactly one winner and one loser. In a well-balanced matchmaking system, such as the one based on Elo ratings, on average, any individual player can expect to win exactly 50% of matches. Since the Elo formula is symmetric, after a certain point, the player's Elo rating is expected to grow as much as it is expected to fall. Thus, the player is stuck hovering at about the same level. To help players break out of this deadlock and progress further in the metagame, we need to break this symmetry. We can do this by adding another factor to the Elo rating update formula,

$$R'_A = R_A + K(S_A - E_A) + L\frac{P_A}{P_A + P_B} + S_A V.$$

The factor V is sort of a bonus that is added to the player's Elo rating with each victory, actually with each match that doesn't end up in a defeat. To ensure this, I multiply it with S_A the variable that is anyway equal to 0 in the case that the player has lost the match. The exact value for the variable V can be chosen freely.

This addition has some important implications for the behavior of the whole system. The maximal value by which R_A can be reduced is $K + L$. However, the maximal value by which R_A can be increased is now equal to $K + L + V$. This means that, on average, all Elo scores for all players will gradually creep up over time. This allows players to eventually progress through the metagame and unlock new content without the danger of being stuck. Furthermore, this rate can be calculated with respect to the number of matches that the player plays, as $\Delta R_A \approx nV/2$ where n is the number of played matches.

But what about the fairness of the matchmaking? Doesn't this break the delicate balance? Actually, what this adds is a certain velocity by which players' Elo ratings creep upwards. If the same V factor is applied uniformly to all the players, the relative differences in their Elo scores will remain roughly the same. Thus, matchmaking will also remain relatively fair.

The really fun part about all this is that you are actually free to play around with the values K, L, and V factors. If you are building a system of arenas or leagues, you can assign different sets of values for these factors for each arena, creating a more dynamic or more static environment, making the actual score have more influence on overall player rating or over-reward victories.

For example, in many league-based systems, the lowest entry-level league, tends to be populated by a mixed crowd of players with very diverse skill levels. You might want to over-reward any victories that a player achieves in this league to make them move to the higher leagues faster.

Social Gameplay

7

Humans as Social Creatures

Sometime about 74,000 years ago, a volcano on the island of Sumatra in what is now Indonesia erupted, ejecting over 2800 km^3 of pulverized rock into the atmosphere. The outfall carpeted the whole of Southeast Asia with a layer of ash 15 cm deep. Graded by our modern scientific scale, this eruption would have reached level 8 on the Volcanic Explosivity Index scale, the highest possible. The shards of molten volcanic glass have been discovered as far as East Africa.

The dust cloud remained in the atmosphere, blocking out the sun. The ensuing volcanic winter lasted for about a millennium. The ensuing climate change caused a massive die-off of large mammals, such as tigers, cheetahs, chimpanzees, and orangutans. The traces of this can be seen in the DNA of multiple species that faced near extinction. Human DNA also indicates that our own species underwent a population bottleneck at roughly this period of time. The population of Homo Sapiens seems to have dwindled down to a mere 3000 to 10,000 individuals. It was a near miss. Each and every of the 8 billion humans on earth today are descendants of these survivors.

It was a near miss, but how did we manage to survive? Well, contrary to popular opinion, hard times, times of cataclysmic disasters, are not the times when selfish people can thrive. What seems to have helped humanity avoid extinction is our ability to cooperate. Only the groups that could stay together and that were willing and able to cooperate and trade with others survived. Others, locked in their own greedy mentality, might have pushed on for a little while, hoarding precious little food and material they could lay their hands upon, but would eventually succumb alone in a vast ashen desert.

This, at least, is what the so-called Toba Catastrophe Theory suggests (Ambrose, 1998). This theory is hard to verify and has been criticized by many sources in later years. It nonetheless tells a very compelling story. What we do know is that we humans are capable

S. Stanković, *Game Design for Free-to-Play Live Service*, Synthesis Lectures on Image, Video, and Multimedia Processing, https://doi.org/10.1007/978-3-031-56156-6_7

of genuine altruism. This may sound like a figure of speech, but the term genuine altruism has a very specific scientific meaning. It denotes a conscious behavior by an individual towards a stranger where there is no direct benefit to the person or hopes and expectations of reciprocity (Vlerick, 2020). We are literally hardwired to seek the companionship of others. Our need to be in contact with other humans, to be embedded in this matrix of humanity, has been identified as one of the fundamental psychological needs. In the Self-Determination Theory (Ryan & Deci, 2020), the psychological framework we already discussed in Chap. 3 uses the term relatedness to denote this need.

This is evident in our everyday lives. Not only do we tend to live in communities, from tiny towns to cities with millions of inhabitants, but we also spend an inordinate amount of time communicating with other humans, from sitting around the campfire to chatting on social networks. Loneliness is one of the greatest challenges to overcome as a human, and solitary confinement is one of the harshest penalties that can be imposed on a person. The human mind starts to crack after periods of prolonged isolation (Wenk, 2023). This is something that we all had the misfortune to experience due to recent lockdowns.

The world of gaming reflects this reality. Multiplayer games such as Fortnite, World of Warcraft, League of Legends, Team Fortress, and PubG to name but a few, dominate charts (NewZoo, 2023). Even the games that are not primarily focused on multiplayer gameplay, such as Minecraft, enjoy a vibrant multiplayer community.

Furthermore, the social aspects of video games transcend the games themselves. People playing games share their experiences with other people online. Twitch streams and playthrough videos on YouTube, Discord channels and Reddit pages, forums, and Fandom Wiki pages are all over our digital landscape, and they are all expressions of our social behavior. They are an integral part of the metagame in the larger sense.

Just as human interaction can take numerous forms, so do social aspects of the games can come in various shapes. In the text that follows, we shall dig deeper into various aspects of social gameplay, especially in its two most important facets, *cooperation* and *competition*.

Competition

We will begin this discussion by examining competition as a type of social interaction in games. The main reason for this is that competition is somewhat easier to comprehend between the two. The goals of the competition are simple to understand: a player, an individual, is facing one or more opponents. The individual's goal is to win. This can be achieved either by defeating the opponents, i.e., by knocking them out of the game, or by reaching some goal faster. The term applied to such gameplay is Player vs. Player (PvP).

This type of gameplay can take many forms. One factor that differentiates different competitions is the number of participants. Competition can be between two opponents. Real-life sports like tennis or games like chess are typical examples of this. These types

Fig. 7.1 Dual mode in Rocket League

of competitions are usually referred to as One vs One (1v1 or 1:1). Clash Royale by Supercell is a prime example of this type of gameplay on mobile. Figure 7.1 shows a screenshot from the duel mode of Rocket League.

In contrast to these are Many versus Many PvP games. Every team sport is such a competition. All the games that employ some type of clan battle belong in this category. This type of gameplay is very interesting as it necessitates some sort of collaboration between team members. I am going to discuss more about the interplay between competition and collaboration in the next section of this chapter.

Finally, there are the games in which a single player competes against a number of other players. Any track and field competition in real life is an example of this. There are games that take inspiration directly from these kinds of sports. Fall Guys and its clone, Stumble Guys, are exactly that: silly virtual obstacle races in which many players take part at the same time. Figure 7.2 is a screenshot from Fall Guys.

The games from other game genres also feature one vs. many multiplayer gameplay. The so-called Battle Royale games are one such genre. These games are a variation of shooters in which every individual player fights for survival against every other player. The playfield is getting narrower and narrower as time progresses to keep the intensity of the fighting constant. The last one standing is the winner. The genre became popular with PlayerUnknown's Battlegrounds (PubG), see Fig. 7.3 and Fortnite is the biggest example currently.

Fig. 7.2 Fall Guys by Mediatonic

Fig. 7.3 A screenshot from PlayerUnknown's Battlegrounds

Battle Royale games also offer gameplay modes in which small teams of 3 to 5 players battle multiple other teams of similar size, making them another type of many vs. many games.

In addition to these examples, pretty much any game in which your score is ranked on some leaderboard against the scores of other human players is a form of one vs.

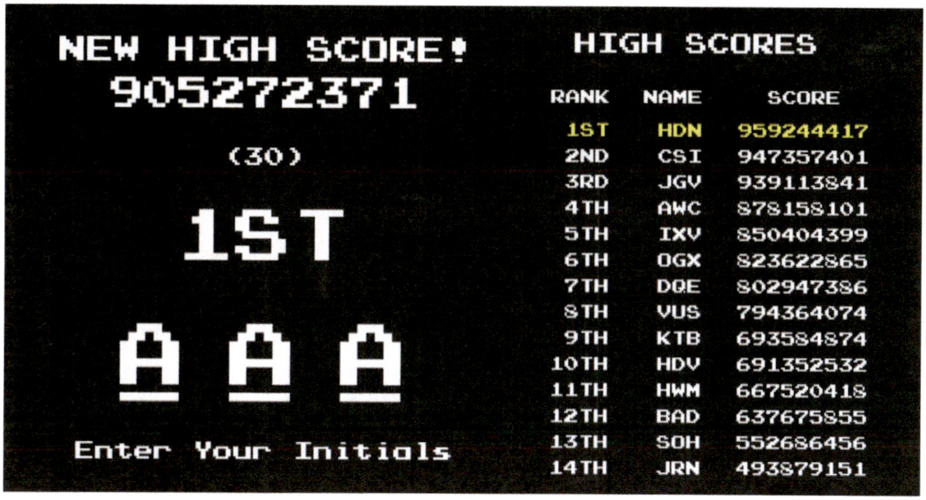

Fig. 7.4 An arcade leaderboard with the three-letter nickname input system

many gameplay. Even in the early days of video game history, arcade machines featured leaderboards. These very, very rudimentary offerings give players a chance to input only three letters as the representation of their identity. Figure 7.4 shows an example of one such system. Yet, to many, it is a big deal to see their three letters rise up the leaderboard on the Pac-Man machine in their local arcade.

The number of participants is not the only criterion that can be used to separate various PvP gameplay models into distinct categories. Players take part in competitions for the joy of competing, i.e. intrinsic motivation coming from the activity itself. However, they also participate motivated by extrinsic factors, i.e. they want to receive some sort of reward as a result of their performance.

The way the rewards are distributed is also an important distinguishing factor. Above all, we have so-called *winner-takes-all* games. In these games, there is only one winner at each round of competition. The other participants are losers and do not receive any prize. A typical example of this is a match of tennis.

These games are a subset of a larger group of *zero-sum games*. In these games, a limited pool of rewards is distributed to the participants, usually according to their relative performance in the competition. Consider for example, in which a hundred players are competing for the rewards. They collect the points in some way and are ranked on a leaderboard. The total pool of rewards is 100 diamonds. At the end of the competition, the highest ranking player, the one in the first spot, gets 10 diamonds, players in places 2 to 5 get five diamonds each, players in spots 6 to 10 get 2 diamonds, and everyone in places between 11 and 50 gets only a diamond. The players who ranked below place 50 do not get any rewards. The total amount of distributed rewards is 100.

Fig. 7.5 A screenshot showing personal goals in a global Truck Order event

This system has one important characteristic. The potential reward that a player might earn depends on his own effort. However, it also depends on the effort of other players, who might push him down in the final rankings.

Other games take a different approach. They distribute rewards based on the player's own effort solely. For example, if a player reaches a certain point threshold, he becomes entitled to a certain reward. The reward thresholds are absolute values and not relative values tied to the performance of other players. Global Events in Hay Day are an example of this model. You can see a screenshot from Hay Day illustrating this in Fig. 7.5.

Finally, the last differentiating criterion that I am going to mention is the role that time plays in various PvP gameplay modes. In quite many of the examples that I have mentioned earlier, including all the battle royale games, Clash Royale, Fall Guys, etc. The competition takes place in *real-time*. In other words, the participants are competing against each other *synchronously*, i.e., they are present at the same time in the same play session.

However, this is not the case in all the games. In many of the games, competition can be *asynchronous*. The participants are able to log into the game at different times. They are able to collect points in their own play sessions independently from other players.

The advantage of real-time PvP is seen already in its name. It offers real-time feedback about the player's and opponents' actions, allowing for fast and dynamic gameplay. This makes it the only viable option for many gameplay genres. On the other hand, it requires enough players to be present in the same play session at the same time for the competition to take place. The async PvP, on the other hand, does not have this requirement. It is better suited for longer, more casual competition.

This distinction between real-time PvP and asynchronous (async) PvP is important, not only from the user experience perspective but also from the technical perspective. For real-time PvP, the infrastructure, i.e., the server backend and the internet connections, need to be able to handle the required amount of information in a very short amount of time. Failing to do so causes problems in latency and synchronization, negatively affecting the user experience and putting some users in a disadvantaged position. The technical requirements are typically much lower in the case of async PvP.

The feature we developed for SimCity BuildIt called Contest of Mayors is a typical example of such asynchronous PvP. In this feature, every city mayor is competing against a group of one hundred other mayors, leading virtual cities of approximately the same level. The competition is a weekly recurring event starting on Monday and ending on Sunday evening. During this time, the participants try to collect as many points as possible. They collect points by completing various assignments around their cities. They are doing this at their own pace in their own cities independently of each other, thus asynchronous. The rewards are distributed at the end of the competition based on the standings on the leaderboard. A screenshot of a SimCity Contest of Mayors leaderboard is shown in Fig. 7.6.

Many games combine the two approaches. Offering fast-paced real-time PvP as the core activity while adding some sort of longer asynchronous competition on the metagame level.

Fig. 7.6 Contest of Mayors leaderboard from SimCity BuildIt

Collaboration

Competition occurs everywhere in nature. Individual animals of the same species compete for food and a chance to mate. Different species compete for living space. The ability to collaborate is much more rare. However, it is an inherent facet of our social behavior. Indeed, it is the main evolutionary driver behind our social behavior. By definition, collaboration means that the individuals within one group are willing and able to coordinate their actions to achieve their goals. We, humans, are also able to exchange resources, help, and comfort each other. They are also able to exchange opinions, discuss, plan, and create strategies.

All collaborative behavior has its roots in social groups. As individuals in modern society, at any given moment in time, we are part of a multitude of more or less formal groups. We have our families. We live in cities and towns. We are citizens of states and other political entities. At work, we are part of a team. Many of us are members or at least supporters of sports teams, etc. All these social groups share many common properties. Our basic social behavior is replicated in the world of gaming. In the video game world, such groups are known as clans or guilds. In this section, we shall examine some of the various aspects of collaboration in the gaming world. But before the collaboration begins, a group, no matter how informal, needs to be formed. This is where we will begin our discussion.

Forming Groups

No story ever starts with "They were a perfect team in which every member knew its place, and they operated as a well-oiled machine". Every story, every epic about teamwork, always starts with a band of misfits. If you ever worked for a big company, you have no doubt been subjected to one of the occasional team-building exercises.

Crude as they are, these corporate rituals try to tap into something very fundamental. Not only that we humans are social creatures that thrive when being in a group, but our groups also have structures, internal hierarchies, patterns of roles, and behavior. This is something we have in common with our closest evolutionary relatives. All primates share this behavior (Swedell, 2012).

Every member of any human group has a place within that group in relation to every other member. This applies to all groups and in all circumstances, from sports teams to corporate boards, including, of course, various teams of players, guilds, clans, clubs, etc., in multiplayer video games. The process of creating this internal structure is what team building is all about. It needs to play out whenever a new group is formed or whenever the circumstances change dramatically enough to affect the functioning of the group in the old format.

Every member of the group assumes a role. There are leaders and followers: the lancer, the foil, the sage, the healer, the grumpy one, the loner, the weirdo, etc. There are hundreds of archetypes that define these group dynamics. The process needs to be revisited whenever a new member joins this group.

The depth and the strength of the bonds between its members create the cohesion of the group. The bonds of kinship, connections with friends we make during our formative years, and the connections with our love partners are the strongest bonds among humans. They are followed by connections with our work colleagues, neighbors, and casual acquaintances.

For a member to stay with the group, two conditions need to be met. First, the member needs to be accepted by the group. Second, the individual needs to be happy with his or her role within the group. This is especially relevant in the world of multiplayer games. People play games to be entertained. They are not compelled to play games to earn a living. They are not bound to reside in games in the same way they are bound to their physical place of residence. New groups are created, and old ones are disbanded all the time in games.

As a game designer, especially if your game is targeting a more massive audience, you are facing two conflicting design imperatives. On one hand, you would like to attract as many players as possible to join the groups that make the heart of your team gameplay. This means that you can't simply count on players forming teams with their friends from real life. On the other hand, you would like to build deep bonds that forge a community and bind players to your game.

You will need to find a way to motivate them to interact and collaborate with strangers. This is in itself a very challenging design task. In order to do so, above all, you need to build a context that provides the reason for the newcomers to engage with the people they do not know and who might be represented only by tiny avatars and made-up nicknames.

Keep in mind that for any newcomer to any virtual environment, the context is always novel. It might be somewhat familiar, but the novelty almost always outweighs the familiarity. In this situation, we as humans resort to a time-tested evolutionary survival strategy. We learn from others. We start to observe what other people do. This is what is known as *social cues*.

The most blatant example of this behavior comes from the world of urban crime. Without a doubt, you gave at least once encountered an example of news about a violent crime that happened in a public place in front of a group of bystanders. The victim was brutalized but no one of the onlookers intervened to help. The blame for this behavior is usually placed on the alienation brought on either by urban life, these things by definition, happen in urban environments, or racial or religious hatred or overdependence on modern technology, etc.

The truth is much simpler. Chances are that none of the onlookers have had any previous experience with similar situations. This was a first-time experience for them as much as for anybody else involved. These things are extraordinary. They don't normally happen.

Nothing in their mental repertoire prepared them for this and they simply did not know how to react. Most probably, they were standing around looking at other equally confused people. As no one did anything, no one else knew what to do. If the first person pulls out a cell phone and starts recording a video, a bunch of other people will do the same.

This is why the first responders, policemen, firemen, various rescuers, and medical specialists go through rigorous training. They are learning from hypothetical scenarios how to react in a situation that they have never encountered before.

Incidentally, if you ever find yourself in a situation where you need help but are surrounded by a confused bunch of gawkers, the way to get yourself out of the predicament is to personalize your request for help. Pick any person from the crowd and ask for anything. Be specific. "Hey, you in the yellow sweater, give me a hand".

The person will most likely move. That one might not know how to help, but soon, someone else will pick up the cue and move in to try to help. Eventually, someone might be able to actually do something useful.

This applies to multiplayer games. New players learn behavior from the environment. Each community creates its own micro-culture, a set of unwritten rules about acceptable and unacceptable behavior. This is why toxic behavior in certain game communities can be so hard to stamp out. The culture of toxicity has already become ingrained in the community. This is also why community management early on is so important. If a healthy culture has been established, it becomes self-reinforcing. Likewise, the bad one can continue to reinforce itself despite the best efforts to weed it out.

Personal Versus Common Goals

Although genuine altruism is a property of the human species, it is rare. It is rare enough that you simply cannot count on it to act as a main motivator for player interaction. We have already talked a lot about the importance of players' goals in game design; see Chap. 5. Multiplayer games add a whole new dimension to this topic. In team-based multiplayer games, we are dealing with two levels of goals: the personal goals of each individual team member and the goals of the team as a whole.

The goals on these two levels might or might not align. This opens a multitude of challenges for a designer, but it also opens a lot of possibilities. Every real-life team sport offers plenty of examples of how two layers of goals might clash. Without a doubt, you have seen examples of star-studded professional football, basketball, or ice hockey teams chock full of players with inflated egos who just can't seem to get their act together and start performing as a team. Well, you got your collision of personal and team goals right there. The team's goal is obviously to win matches, but the personal goals of individual players are something entirely else: to look good at prospective against that might offer a more lucrative contract for the next season in another team, to push up some stat number or to go for the personal record, or to simply be the center of attention.

The beauty of video games is that they have more freedom than professional sports. Things that are generally detrimental for a sports team might be interesting in the game context. Of course, the strongest interaction between players happens when personal and team goals align. The secret is again to offer a plethora of both the personal and the team goals to choose from and thus increase the probability of this alignment.

Another challenge is the difference in abilities between players. A successful game should be attractive to as broad an audience as possible. This, in turn, means that you can expect various players will have a very broad range of skills, from total noobs to fanatics demigods that eat, drink, and breathe your game. Their ability to reach goals will be proportional to their skill. In order to keep all of them interested, you need to provide a range of goals that scale with their ability.

You can also provide a set of tools for players to select the groups that they would like to join but correspond to their skill level. You should also provide a set of tools for clans to decide which players they want to accept. In effect, you need to reveal some indicators of the player's skill levels. This could be a trophy count, the player's XP level, a set of badges that can be earned only in a specific way, etc. However, there are several pitfalls when designing these features. Putting too much emphasis on skill might create an overly elitist community culture that would inevitably push your game out of the broad audience and towards a gaming niche. Anecdotally, joining some of the top corporations (guilds in their parlance) in EVE Online is harder than landing a job in some of the top companies (Yancy, 2019). The financial success of EVE Online shows that targeting a relative niche can be a very lucrative strategy.

On the other hand, you might want to employ a more elegant method and allow players to choose goals that they can pursue that are commensurate to their skill level. If every player can pursue his or her own private little goals and still contribute to the success of the group, you are onto something powerful. If you manage to pull it off so that even the most novice player can have a meaningful contribution to the success of the group while simultaneously pursuing personal gains, you are on a very good track.

Shallow Interactions

Of course, this trick is very hard to pull. A lot of game designers seem to forget this and focus either on the personal goals of players or on global group goals. This can often result in very shallow social interaction.

Typical examples of this are various gift-giving or helping mechanics that were prevalent in farming games during the Facebook era. For example, in the original Farmville or indeed even in Pokemon GO, each player is able to send a limited number of gifts to other individual players in the game. There is no overarching group goal to this. There is no cost to the player sending a gift, and there is usually only some marginal benefit for the receiver. Usually, a gift contains a handful of useful but relatively common items.

Such interactions were meant to benefit from another feature of the human psyche, known as the principle of *reciprocity* (Fehr & Gächter, 2000). The principle of reciprocity is another evolutionary survival strategy. This principle states that we tend to reply to a positive action with another positive action and to a negative action with another negative action. Most probably, you are familiar with the concept of a *prisoner's dilemma* (Rapoport & Chammah, 1965). It is a clever little concept postulated in mathematical game theory. I will retell it here for the sake of completeness.

Imagine that you are one of two gangsters members of the same gang. Both you and your buddy have been arrested. The cops are keeping the two of you locked in separate cells. The inspector claims that they have enough evidence to put you behind bars for one year. However, the prosecutor offers you a bargain. If you agree to testify against the other guy, he will get three years in the slammer, and you get to go home scot-free. But here is a kicker. You know that they are going to offer the same deal to your buddy. If both of you decide to cooperate, both of you will end up in jail for two years. Now, you ask yourself what should you do. Keep your mouth shut and hope that the other guy will do the same and end up spending a year in the big house. But what if he snitches on you? You'll end up spending three years behind bars, and that bastard will walk free. You could risk it and rat on him, but if he does the same, you both will end up two years in jail, so what did you do? You just made things more difficult.

Formal mathematical theory treats this as a paradox. There is no perfect solution to this conundrum without involving psychology. To get out of this situation, you simply need to trust the other person, or at least know what the other person is likely to do. The reciprocity principle is the prisoner's dilemma looped multiple times. Each instance of interaction with the other person gives us knowledge about its probable reaction and behavior. This apriori knowledge helps us make a decision in the prisoner's dilemma. If someone mistreated us once, we are likely to expect the same and act accordingly. If someone shows us kindness, we are likely to respond in turn.

The gift-giving mechanics, as implemented by these games, were meant to reinforce bonds between players using this method, i.e., by artificially motivating players to engage in this sort of reciprocal behavior. The problem with this implementation is that there is no real emotional value to this interaction. As there is no cost to the sender, the receiver feels no value in the gesture and does not feel obliged to return a favor. The system instead relies on rewards, i.e., on external motivation, see Chap. 3, to try to motivate the players to interact with it. Send gifts to Marco ten times to gain another heart. Congratulations, your friendship level with Bob is now 3. The process becomes quickly, strictly mechanical, and transactional. What was supposed to be an engagement mechanic through social interaction turns on itself and starts to rely on classic gameplay tropes to maintain its own level of engagement.

The secret to avoiding this trap is to require some real effort from players when interacting with others. To justify this demand, allow them to simultaneously pursue their own selfish goals. Altruism and selfishness are the yen and yang of human nature.

Deeper Interactions

As we mentioned earlier, due to the reciprocity principle, the more effort required, the deeper the bond between the participants in the social interaction. This is well illustrated by the examples of various raid systems in many multiple games. Raids are primarily player versus environment type of gameplay. They are typical collaborative features. A group of players is required to collaborate in order to defeat one gigantic monster or reach the end of the dungeon. This type of feature is a standard in many MMO RPGs, from the World of Warcraft to Guild Wars 2 and Final Fantasy XIV. In games like the Monster Hunter series, they are the centerpiece of the gameplay. They are common even on mobile Free-to-Play RPGs such as Summoners War but can also be found in a simplified form in games like Pokemon GO. Figure 7.7 shows an example of a raid in Summoners War.

I am using raids as an example because the goal structure is usually extremely easy to understand. The group's goal is obviously to win, to successfully complete the raid, for example, to kill the boss monster. The goal of individual players is to gather as big of a reward as possible out of the raid. It is obvious that these two levels of goals align. In some games, the player needs to survive the raid in order to partake in the sharing of rewards. In most games, the amount of rewards that the player gets is proportional to the player's effort. For example, the more damage the player deals to the boss, the bigger reward the player gets.

These two layers of goals align perfectly. As each player tries to maximize his or her own payout, they quicken the victory for the whole group.

Fig. 7.7 A raid in Summoners War

Once again, the amount of effort dictates the depth of social bonds. Some games employ very simplistic raid battle systems. In Pokemon GO, for example, players are not required to coordinate their efforts very much. It is enough to strike at the boss with the most efficient Pokemon in the player's selection. Other games on the other hand, require much more careful and intricate strategies.

The main design challenge here is that not all players are willing to invest the same amount of effort to achieve the goals. If the raid system is too demanding, many players will opt out. If it is too forgiving the depth of social interactions will be too shallow. The trick is finding the sweet spot.

Conflicting Interests

While working on SimCity BuildIt, we constantly tried to expand its social gameplay. The game started off as a pretty solitary experience, offering a pretty rudimentary interaction between players. One idea we constantly encountered while searching for inspiration was about building a city collaboratively. It is a game about building a city. Why don't we let players build a city together? It sounds natural in a way, but if you ponder it even a little bit, you can realize that it is actually a terrible idea. The basic premise of SimCity is that it puts the player into the shoes of a mayor. Actually, not only a mayor but a mayor, an architect, and a chief urbanist at the same time. It allows the player to live out a fantasy of having ultimate control over a vision of a perfect city. This is what makes the game compelling to the player, but it also implies that there is only one mayor whose vision gets to be executed.

If multiple players are to collaborate in building a city, whose vision is it going to be? Are they all going to try to build their own cities of their dreams in the same spot? How to avoid the city becoming a battleground of competing visions? Is only one player going to have the privilege of deciding about how the city should look, and everyone else be confined to some subservient role? Or is there going to be no vision at all? Isn't the city going to devolve into a pseudo-organic mess? If there is no fantasy of creating a vision, could there be an emotional connection, or would the whole thing devolve into a mechanistic gameplay requiring external motivators to work?

We never managed to get the whole thing working. The potential conflict of interests of various players was just too great. This is another important dimension to the discussion about goals. You need to keep an eye on the goals of each individual versus the goals of the group, but that is not all. You need to be mindful of how the individual goals of various members of the group interact with each other. Do they align? Are they conflicting? Is this conflict part of the gameplay?

On the other hand, a project known as Reddit r/place offers an example of how this mass collaborative effort may succeed in producing something coherent and compelling (Rappaz et al., 2018). It is not a game in the common sense of the term. More accurately,

Fig. 7.8 The 2023 version of Reddit r/place

it could be described as an art project that has many gamelike elements. This is a recurring yearly event. It is a blank virtual canvas of 1000 by 1000 pixels. Each registered user can place pixels of any color anywhere on the canvas. However, only one pixel can be laced at a time. A certain amount of time has to pass before the same user can place another pixel. The duration of the whole project is also limited, putting a hard cap on the number of pixels each individual user can place. As human minds abhor chaos, users obviously try to create something recognizable using the means at their disposal. They can choose to collaborate to speed up the process, but various groups for the limited space on the canvas. The result emerges organically from this interplay of collaboration and competition. You can see the results of the 2023 event in Fig. 7.8.

Collaborate to Compete

As we have seen, collaboration alone can offer compelling gameplay. However, a truly potent mix arises when collaboration is mixed with competition. The interplay of these two aspects of our social behavior offers endless possibilities for fun and engaging gameplay that creates deep emotional bonds.

We have already mentioned one example of collaboration in a competitive setting in real-time PvP, this of teams in battle royale type of games. Of course, if you are at

Fig. 7.9 A screenshot from Team Fortress 2

all familiar with the world of gaming, you know that these examples are much more numerous. The battle royale games themselves owe much to an older genre of team-based FPS games such as Overwatch and much older Team Fortress 2, see Fig. 7.9. This genre is a natural evolution of the single-player FPS games that evolved on the PC in the middle of the 90 s.

Psychological analysis has shown that this particular genre has a great ability to satisfy all three basic psychological needs as identified by the SDT framework: agency, mastery, and especially important for this discussion relatedness (Rigby & Ryan, 2011). For more information, see Chap. 3.

However, this is not the only genre showing these properties. Another good example of collaboration during competition in gameplay is Multiplayer Online Battle Arena (MOBA) games. These are third-person games in which teams of three to five players battle against each other on a predefined map. These games evolved from an earlier real-time strategy genre and became popular with titles such as Defense of the Ancients (Dota), its sequel known as Dota 2, see Fig. 7.10, and League of Legends.

This genre offers the same potent mix of psychological drivers as team-based shooters. There are numerous examples of team-based multiplayer gameplay in the mobile Free-to-Play world. Indeed, many of the titles I have already mentioned are multiplatform games playable on both iOS and Android.

There are also numerous other native, purely mobile games that are built around this type of gameplay. For example, Brawlstars is one such game. Its gameplay has many

Fig. 7.10 A screenshot from Dota 2 MOBA game

common elements with both of the previously mentioned genres, combined with many unique elements. A screenshot from this game can be seen in Fig. 7.11.

Fig. 7.11 Brawlstars by Supercell

Fig. 7.12 EVE Online by CPP

The three previous types of games, team-based FPS games, MOBAs, and brawlers, are all focused on small teams of players. Massive Multiplayer Online RPG games take a different approach in these virtual worlds, thousands of players can be members of the same guild or clan. A great example of this is corporations in EVE Online, see Fig. 7.12.

All of these examples are focused on real-time gameplay. However, the asynchronous multiplayer modes also offer many great examples of a combination of collaborative and competitive gameplay. An iconic example of this in the mobile world is Clan Wars in Clash of Clans by Supercell. In this gameplay mode, two clans of players fight each other. The war is composed of a series of individual battles and attacks of individual clan members on the basis of the members of the opposing clan. These attacks happen asynchronously from each other, and only the final tally determines the winner. Figure 7.13 shows a map of one such clan war.

Finally, the whole genre of so-called 4X games hinges on intense social gameplay. These games are an evolution of the strategy genre with a heavy emphasis on multiplayer gameplay. The core appeal of these games is rooted in social gameplay, where players are expected and heavily motivated to forge alliances, share resources, and strategize in order to compete against other players engaged in similar activities. This genre is typified by Game of War Fire Age by Machinzone.

Fig. 7.13 Clan War map from Clash of Clans

Competition Within the Group

Another way in which collaboration and competition interact is a competition that can take place within the group. As we mentioned already, all human groups have some internal structure. Very often, this structure has some form of formal and informal hierarchy. Whenever a hierarchy emerges, some individuals will instinctively try to improve their position within it. Even in the case that the game does not offer explicit means of competition between individual group members, this behavior can take place in canals of communication like in-game chat or elsewhere.

Many games offer more explicit means to establish hierarchy, such as implementing a series of ranks that players can have within their clans or guilds. There could be a system of badges that a player can earn for specific achievements. In other cases, general player properties, such as player XP level, will serve a similar purpose. Clan leaderboards are also one very common example of this type of mechanics.

Furthermore, in many cases, the system in which rewards of collective activities are distributed is a zero-sum game, where a finite number of items is distributed to the participants based on their contribution. This can be a very harsh mechanic as more skilled players can get the lion's share of the winnings, while others who were not able, for whatever reason, to contribute might feel excluded.

Really interesting group dynamics start to evolve when formal and informal hierarchies of the group do not align. For example, some members of the group start to feel that they have too little influence on the group's decisions compared to what their formal status implies, and vice versa.

This is, of course, very common in all human social interactions, and this is a perfect illustration of my initial observation, see Chap. 1, that games serve as little laboratories for human learning. Soft skills acquired while dealing with interpersonal problems within an MMORPG guild are directly translatable to the other spheres of life (Pagel et al., 2021).

Identity and Social Status

We might be social creatures, but we are also individuals. Even as members of the group, we do not wish to lose our identity. Quite the contrary, we have a deep-seated desire to be recognized as distinct members of the group, to define and assert our identity. Human society has invented many ways in which we accomplish this. Wearing clothing and personal jewelry is something shared by all human cultures. Wearing particular kinds of clothes or items of jewelry can offer much information about our place in society.

Universities often pride themselves as unconventional places, not burdened by the strict dress codes and formalities. I completed my studies at three universities and visited many more during my time in academia. After a while, I noticed that all of these places shared a secret informal but rather strict dress code. All of the undergrad students would be dressing rather casually in t-shirts and jeans, and as they advanced to their doctoral studies, t-shirts would be gradually replaced by shirts and sweaters. On the postdoc level, casual smart sports jackets would start to prevail. Younger professors busy with chasing grants would tend to wear full formal suits akin to ones worn by business people and lawyers. Finally, older tenured professors nearing retirement would gradually switch again to casual, smart combinations of sweaters and blazers. Breaking this dress code, by god forbid, dressing in the garbs of the wrong rank, would inevitably attract stares in the school cafeteria.

In more obvious and banal examples, various professions wear distinct uniforms to be recognized as firemen, policemen, doctors, nurses, etc. We chose items of clothing to stand out. Sometimes, to indicate the function that we serve in society. We do the same to fit in. Have you ever felt overdressed for the occasion? Have you ever shown up dressed casually for a formal event cos you missed the memo?

Sometimes, we dress to fit in to stand out. For example, if all the cool kids in your school were goths, and you wanted to fit in with this cool crowd, you'd dress yourself appropriately so that everyone at first glance could identify you as a member of this elite group.

What we wear can indicate our social status. The insignia on the general's shoulders indicates his rank. An expensive Rolex projects power and wealth.

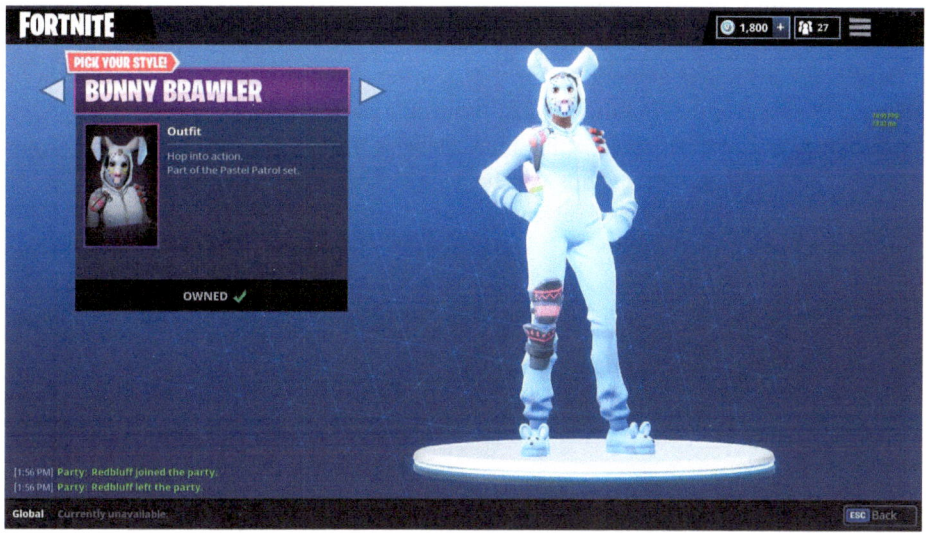

Fig. 7.14 Bunny Brawler from suit from Fortnite

Naturally, all of these aspects of human behavior have their counterparts in the world of video games. The category of virtual objects that serve this purpose is known as *vanity items*. Typically, they can be obtained either by playing the game or via direct or indirect purchase.

These vanity items can take many forms, typical ones being various skins or suits that change the appearance of the player's avatar or character and various items of clothing or gear that do the same but offer more options for micro-customization of the character. This is typically seen in team-based shooter games. Figure 7.14 shows an image of the famous bunny suit from Fortnite. In addition to this, many games offer things such as celebratory dance moves.

In games that do not feature such prominent avatars, vanity items can have other forms. Such as various flags, emotes, and even decorated frames for the player's image.

However, to serve their overarching purpose of establishing a player's identity and showing his social status, two criteria need to be met. Above all, an audience to see them needs to exist. This is why these items make very little sense in purely single-player games. If there is no one to see your fancy clothes, why bother? You can continue to dress like a hermit. This is what everyone was doing while we were in our goblin mode during the pandemic. Second of all, items need to be exclusive. A Rolex is a Rolex because it is expensive. Only someone promoted to the rank of general can display that insignia on the uniform.

The availability of vanity items in the game needs to be subjected to virtual scarcity. The most valuable items need to be either hard to obtain, expensive to purchase, or available only for a limited amount of time. It is often helpful if some kind of anchoring value can be established. In other words, ostensibly cheaper items should exist to offer a contrast and reinforce the value of the more exclusive items.

The importance of vanity items can be seen in their monetization potential, as they constitute a significant chunk of revenue for many Free-to-Play games.

Game Design Methods

8

Game Development Process

I remember a period of time of about four years while I was working at one of the major game development companies. During this time, one midlevel manager used to come over from the central office to our studio about once a year. On the company level, this person was in charge of developing an efficient product pipeline specifically for the development of new mobile games. Each year, this person would appear with a new PowerPoint deck with a new product approval and development process. Every year, a brand new chart. Every year, the same person.

Mind you, this was not an isolated case nor a quirk of that particular company. It was just the most obvious example of the thing I witnessed over and over in pretty much every place I ever worked with. I am telling you this because I want to underline one simple fact: the games industry has not yet come up with a single standardized recipe for how to bottle a lightning.

Goals of the Process

Making games is hard. Making good games is harder. I don't know if I ever made a good game but I do know one of the dirty secrets of the games industry: there is no single game development process! This process differs not only from company to company or from studio to studio but, damn well, from game to game. On the other hand, this doesn't mean that all of these approaches to game development are completely and totally different to the point that would make them totally unrecognizable. Quite the contrary, they do share quite a lot of commonalities.

© The Author(s), under exclusive license to Springer Nature Switzerland AG 2024   153
S. Stanković, *Game Design for Free-to-Play Live Service*, Synthesis Lectures on Image, Video, and Multimedia Processing, https://doi.org/10.1007/978-3-031-56156-6_8

Above all, all of them share a common set of goals, which can be broken down into two broad categories:

1. Coming up with a suitable game idea and evaluating that idea from various perspectives in order to minimize the risk of development,
2. Actually, implementing the game with available resources in a reasonable time while hitting an acceptable quality level.

It is important to notice that there are also two perspectives of looking at these questions:

1. From within the team,
2. From the perspective of external stakeholders.

This particular section of this chapter will focus on the team's perspective. Depending on the type of the company, external stakeholders might be different. In an independent startup, the external stakeholders might be investors of different sorts, i.e., venture capitalists, investment funds, angel investors, Kickstarter supporters, etc. In the case of studios that are part of bigger companies, external stakeholders include the people in charge of the game portfolio and the pipeline at the company level. These groups of people might have different perspectives on the questions that we are discussing. In general, however, they will try to minimize what is known as the opportunity cost of game development. This is something that we are going to discuss in the next section of this chapter.

In general, in order to reach the two goals stated above, we need to be able to answer two deceptively simple questions: Why? and How? The first of these questions expands into a set of interconnected inquiries. Why should we make this game? Who is it for? Who is going to play it? Does the world need another game? Why is it fun? In the big corporate lingo, this line of inquiries is often summarised under the label of *product fit*.

People looking from the outside into the games industry, as well as people making their first steps in the industry, tend to fixate on the notion of game ideas. Unfortunately, another dirty secret of the games industry is that ideas are easy to come by. They come from all possible sources and from every possible direction. What is scarce are development resources and especially development time. Choosing which ideas to focus on is difficult. We shall discuss various perspectives on evaluating ideas in one of the following sections of this chapter.

The second question, the question of How? is one about the *feasibility* of the game development. It, too, hides a series of sub-questions. Do we have the technology needed to develop this game? Do we have the right team to build it? Do we have all the necessary resources? Can we build it in a reasonable amount of time? Can we achieve the needed quality level? ...and so on. These questions are often depicted in the form of a so-called production triangle, where each of the corners of the triangle represents one of the facets of the production feasibility, as shown in Fig. 8.1.

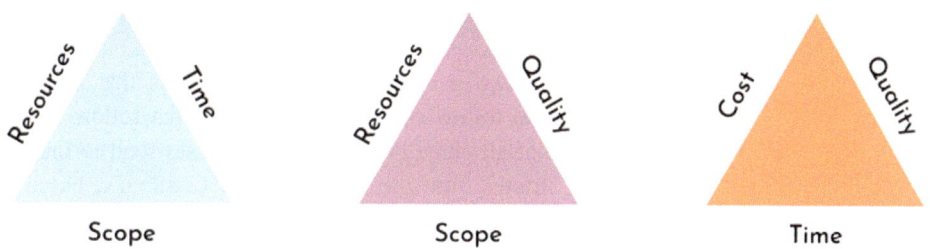

Fig. 8.1 Various forms of production triangles

In a well-organized and structured design team, the process of finding the answers to these questions needs to involve every team member. Specifically, the key stakeholders, such as for example the design lead, the art lead, and the technical lead, need to be able to provide their input on the key aspects of the game development process. However, the various domains of each discipline should be respected. A dedicated owner of the production process itself is usually required. By owner, I mean to say there should be at least one person taking on the responsibility of making a final judgment about the feasibility of the undertaking. In my experience, it is best if this person is neither one of the previously mentioned design, art, or technical leads.

Specifically, the design lead shouldn't be the one making calls on the feasibility of the development in order to avoid two specific pitfalls. The pitfall number one is having no restraints on the design ambition, totally disregarding the time and resources needed to execute the designer's vision. This is the bane of so many indie game projects as well as the famous vaporware titles of which the history of video games is so full (Draven, 2020). The second pitfall is one of self-censorship, where the designer will go in the opposite direction, restraining the ambitions so much as to provide just the bare minimum of the design to the detriment of the product itself. In a well-balanced team, the design and the art lead should constantly ask, "Could we?" the tech lead should answer, "Yes, but only if…" and the producer should say, "No within this timeframe". If the conversations are flowing differently within your team or if the roles are mixed, you might be running a systemic risk to your project.

Finally, these two main goals of the development process are interdependent on one another and often entangled. Quite often, the game development team needs to revisit them multiple times during the process. This is not uncommon. It is not something inherently bad or dangerous. Quite the opposite, regular checkups can be healthy for the team. We are going to discuss this matter in detail in the final section of this chapter when we examine the m, metaphor of the roadmap. However, keep in mind that the particular topics of your conversations should evolve. If your team is forced to go back and discuss the same design or implementation details over and over again, ad nauseam, this is probably another sign of a deeper systemic problem!

The Process Itself

As I mentioned before, there are no two game development processes that are quite the same. However, they all do tend to follow similar timelines. In what follows, I will describe some of the usual phases that all game development processes tend to include. Note, though, that I am speaking strictly from the perspective of GaaS, i.e., based on my experience with mobile Free-to-Play games. The development process for premium games, especially games for consoles and PC platforms, differs significantly.

I will also focus on the development process from the team's perspective. The influence of the external stakeholders on the game development process usually takes the form of regular milestone checkups, at which the development team is supposed to provide answers to specific questions. These questions can differ very much from situation to situation. They depend on who the external stakeholders are and on the stage of the development process. However, most importantly, the questions depend on the level of trust between the team and the external stakeholders. The more trust the team enjoys, the fewer the questions and the milestones there will be. The frequency of the milestone checkups is a good proxy of the confidence that the development team enjoys!

In general and on the highest level, the development process of games that are intended to run as a service goes through four main phases, as shown in Fig. 8.2:

1. Ideation
2. Preproduction
3. Production
4. Live Service.

As we shall see each of these phases includes some sort of validation of the work done during that particular phase. We can make a somewhat more detailed timeline by including a couple of these validation steps, as shown in Fig. 8.3. Furthermore, each of these phases comes with a set of deliverables expected as its outcome.

Phase number one of the game development process focuses almost entirely on the first main goal of the development process, i.e., on the question of what kind of game the

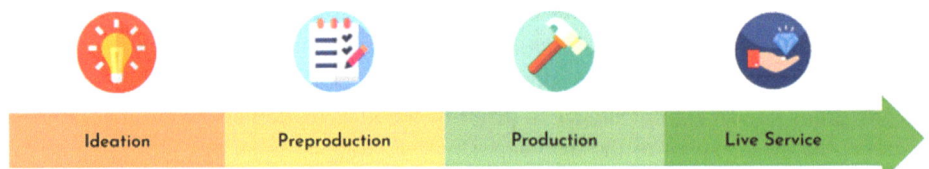

Fig. 8.2 Four phases of the game development process

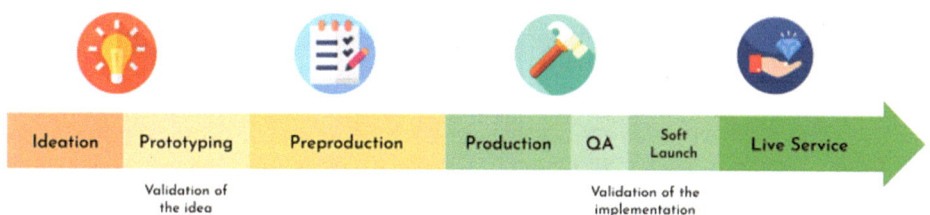

Fig. 8.3 Game development process, including the validation steps

team is making, what it is supposed to be, who is supposed to be playing this game and why should the team make it in the first place.

Ideation

This phase obviously begins with the raw game idea. However, it goes well beyond this. First of all, the raw idea needs to be flashed out in more detail. This is when the idea evolves and becomes the actual game concept. At the end of the first step of this phase, you should be able to provide an elevator pitch to your game, i.e., you should be able to summarise your idea or ideas into two to three short sentences. Treat this as a deliverable. If you are unable to create such a short summary of your game, you are still dealing with a rough set of vague ideas that still need to be flushed out.

On the other hand, if you are able to provide an elevator pitch for a game that makes sense to other people, your idea is ready for the next step of the ideation phase, the concept. In this step, the idea gets more meat around its bones. The idea becomes a concept when more design details are added. There are two things that are expected as an outcome of this particular design step:

1. the outline of the core loop of the game,
2. the list of the things that need to be prototyped.

The core loop of your game should obviously focus on the minute-to-minute interaction of your game, i.e., on its very core gameplay. However, it should also include a reflection on how core interaction integrates with the potential metagame. Figure 8.4 shows an example of the core loop of Clash Royale by Supercell.

The second important thing that you should strive to achieve during the concepting step of the ideation is to come up with a set of questions that need to be answered during the prototyping of the game idea. Prototyping constitutes the validation step of the ideation phase.

Prototyping can take different forms. Sometimes, it can be done in the form of a paper prototype. Often it includes some level of implementation in the code. Prototyping has its

Fig. 8.4 Core loop of Clash Royale by Supercell

own important purpose. It needs to clarify as many as possible of the unknowns tied to the game idea. Sometimes, these unknowns are related to the minute-to-minute interaction. These kinds of things are usually the easiest to clarify with the interactive prototype done in code.

Typically, the team involved in all steps of the ideation phase will be very small, including, for example, one game designer, one programmer, and perhaps one artist. It is not uncommon that at this stage, a single person takes over more than one task, a designer providing art assets, for example, or working with the code.

Sometimes, prototyping involves exploring the technical feasibility of the design vision. In this case, the team composition will differ from the typical and include other developers with a specific set of skills, such as, for example, backend developers, technical artists, shader programmers, etc.

Once you obtain a prototype, you need to actually put it to test. There are a whole slew of methods that are employed by development teams. Internal playtests within the development team is by far the most common one. It is extremely cheap and usually very quick to organize. All you need is a stable enough prototype and some time to gather several team members in a common space to play the game and discuss it. Keep in mind that the participants in this test are usually people very acquainted with the game and the game development process. Their feedback will be extremely biased!

This can be somewhat mitigated by various forms of external game tests, from ad-hoc tests with friends and family members to formal focus group tests. The further removed from the development process the subjects are, the more objective their feedback will be.

In addition to the gameplay perspective, a game idea usually needs to be evaluated from several other angles. Technical feasibility and market potential are two of the most

Table 8.1 Summary of the ideation phase

Steps	Team	Deliverables
• Ideation	• Game Designer	• Elevator pitch
• Concepting	• Game Artist	• Core loop
• Prototyping	• Client Programmer	• Playable prototype

important additional aspects of this. We are discussing the topic of idea validation in detail in one of the following sections of this chapter.

Finally, Table 8.1 presents the summary of the ideation phase of the game with its steps, team composition, and expected deliverables.

Keep in mind that at this stage, your project is still dealing with a tremendous amount of unknowns. The ideation phase of the project is highly iterative. You need to be able to quickly modify the current design ideas or come up with new ones, elaborate them as a concept, and test them out in any way possible!

The general outcome of this first phase of the game design process is that the game idea has been flashed out, tested, and validated. Furthermore, the team has a high-level understanding of what kind of game they are trying to make.

This is simple to understand in theory. In practice, however, there are plenty of cases when the design team has failed to do this and yet sleepwalked into the next phase of the project. There are two main reasons leading into this trap. As we pointed out, the design process at this phase of the project is highly iterative. As with any iterative process, the key question is when to stop with such iterations. The obvious natural choice for this is to stop at the moment when the design is not improving anymore. This, of course, can feel arbitrary. Each iteration brings on some change, but when does a change stop to be an improvement? On the other hand, the iterative process cannot continue indefinitely. Faced with the time pressure, a team might decide to stop the process before the game idea has been properly validated and slides into preproduction in the vain hope that things will somehow eventually sort themselves out. They won't. This simply never happens. It is a recipe for disaster.

Preproduction

The second phase of the development process is the one in which an idea evolves into a real and proper game. The preproduction is the phase in which the vision of the game takes form.

In the case of Free-to-Play games, this includes multiple aspects. Above all, it includes the basic outline of the game's metagame. It also includes the vision of how its live service is going to be operated. Any metagame exploration, prototyping, and testing needs to happen during this phase of the development. At this stage, the design of the game needs to go beyond the gameplay. Things such as branding, narrative elements, character design, world-building, and art style need to be decided in this phase.

Again, some aspects of this process are going to require several iterations. However, keep in mind that further into the development process, you should have more under-standing of the game you are making and should need to iterate less. It is usually a very bad sign if, at this stage, you discover that you need to revisit some of the core aspects of the gameplay.

Furthermore, the feasibility of the execution of this vision needs to be discussed. This includes the technical aspects of the implementation, server, and client architecture, as well as the art asset pipeline. In order to find the answers to these questions, the team working on the game needs to be expanded to include the discipline leads and other key roles within the team.

Finally, a clear implementation roadmap should be an outcome of this phase. This includes the scope of each following major implementation milestone. The scope includes the list of necessary features and the ideal amount of content needed for the subsequent milestones.

The notion of development milestones represents an evolution of the process from the days of premium games on consoles and PCs. In general, any feedback that a team can get from internal testing and even from the focus group testing is going to be biased and of limited use. The only real feedback about the quality of the game you are making can come from actual players, i.e., from the ultimate arbiter of the global audience as soon as you get this feedback, the better. Thus, the idea of release milestones was born.

In the olden days, teams would develop a game in a vacuum, behind closed doors in dark, arcane cavernous laboratories, and release it to the world in its perfect form. Somewhere, by the late eighties, this proved to be insufficient to ensure the quality of the games, and an idea of the beta and even alpha release of partially finished games gradually evolved. Of course, in the times when a game could be released only on physical media, this was a major stumbling block. No one would go through the trouble of stamping and distributing millions of copies of a game that is clearly a work in progress. Digital distribution was a game changer in this respect, and various open betas and early access games are nowadays a norm in the worlds of premium games on both PCs and consoles.

In the mobile Free-to-Play world, this evolved into the idea of a soft launch. We are going to discuss the importance of the soft launch very soon. Therefore, a minimal set of release milestones includes a soft launch release and a global launch release of the game. However, please note that in recent times, these two releases have not been enough. Many, if not most, game studios now include several pre-soft launch releases.

Finally, if you are aiming at the serious success of a game as a live service, your plans should include a roadmap and the feature scope for at least the first several months to a year of live operations after the global launch of the game.

Note also that your game should not contain all the planned features even at each milestone, not even at the global launch. A live service game is in the worst shape at

Table 8.2 The summary of the steps, required team members, and expected deliverables of the preproduction phase

Steps	Team	Deliverables
• Metagame concepting • Technical feasibility assessment • Art style testing • Theme exploration • Narrative outline • etc.	• Game Design Lead • Art Lead • Tech Lead • Product • Manager/Business Lead • Producer/Development Lead • Backend developer • Concept artists • etc.	• Product vision • Art style • Theme, world setting, and narrative • Metagame outline • Roadmap • Scope of milestone releases

the time of its launch. The game needs to evolve and find its audience, and you, as a game designer, need to learn from the game's audience. Allow yourself some time for this learning. Focus on a minimal set of features that the game needs at each stage of its evolution. If the core gameplay is not able to stand on its own legs, no addition of other metagame features will make it win a 100 m sprint.

Table 8.2 contains the summary of the preproduction phase of the game.

Production

Arguably the distinction between the next two phases of the game development process is somewhat blurred when it comes to Free-to-Play games. The production of new features and especially new content for the live service game never actually stops. However, we are going to talk here about the production phase in its narrow sense, i.e., until the game is released to the general public.

As the name implies, this is the phase of the process when the bulk of the game gets implemented. As we pointed out before, this phase of development can include several release milestones, the most important being the soft launch milestone and the global launch of the game.

The purpose of the soft launch is to learn as soon as possible about the game from the actual live audience. These learnings can include tests of the technical aspects of the game, such as stress tests of the game backend, as well as gameplay-related learnings, typically about the balance of the metagame, monetization potential of the game, etc. The idea, of course, is to use these learnings to improve the game before the major global release.

Traditionally, the teams working on Free-to-Play games performed the soft launches by releasing the game either in a limited geographical location on a limited number of platforms or a combination of both. Several smaller markets served as proxies for other bigger, more important markets. For example, places like Australia, Canada, and New Zealand served as proxies for the US, Thailand served as a proxy for Japan and South Korea, and Hong Kong and Taiwan served as proxies for mainland China.

Many game studios released the game initially only on the Android platform, mainly because this platform includes a wide variety of low-performance devices offering the opportunity to better test the game's technical performance.

Releasing a game in a proper soft launch involves building a significant number of features and the amount of content, usually the core game and the basic outline of the metagame with enough content to last players during the envisioned soft launch period. This, in turn, implies a significant investment in development resources and time. In order to minimize this risk, many studios nowadays opt for so-called Limited Market Tests (LMTs). These tests usually involve releasing a very early version of the game to a relatively small number of users. The number of players that get access to such a game is controlled by the amount of money spent on the user acquisition during the playtest and can include anywhere from 10 to 100 of thousands of players. The objective of these tests is to additionally evaluate new game concepts before the actual production takes place. These LMTs, strictly speaking, are a part of the preproduction phase of the development process.

In addition to this, there can be several iterations of the game during the soft launch. New features can be gradually added to the game, or the design and the implementation of various existing features can be done.

Each of the releases during both the preproduction and the production of the game needs to go through a more or less thorough *quality assurance* (QA) process, as indicated in our diagram in Fig. 8.3.

The soft launch period of the process can vary very much in length. There are examples of games that have been released after only a couple of weeks in soft launch to games that have spent more than three years in it.

The outcome of the soft launch can be the final verdict on the game. In a way, this is the final validation of the game idea, the concept, but also of its execution and implementation. Many games have been killed in this period if the KPIs have not met the preset criteria. The KPIs that play a part in this part of the process are, in general, the ones related to engagement and retention, such as D1, D3, D7, and D30 retention. In the past, monetization KPIs also used to be a part of this evaluation process. To my understanding, this is still the case in many companies. However, many others chose to ignore the monetization aspects of the game during this phase and focus solely on player retention.

The development phase of the project bz definition includes the full development team. The composition of the development team and its size depend highly on the type of game and the project scope. The final outcome of this phase of the development should be a vertical slice of the game or the *minimal viable product* (MVP), i.e., a minimal set

Table 8.3 The summary of the production phase

Steps	Team	Deliverables
• Limited Market Tests • Soft Launch • Global Launch	• The whole game development team	• Vertical Slice • Minimal Viable Product

of features required for the global launch of the game. We provide a summary of the production phase in Table 8.3.

Finally, the live service phase of the project includes the part of the game's lifecycle that takes place after the global launch of the game. This is the topic that we are discussing in great detail in Chap. 9 of this book.

While this process applies to launching totally new games, it is important to note that making new features for an existing live game can also follow a very similar trajectory, be it in a very condensed form. In other words, here too, the design will start with ideation, moving on to concepts and quite often, prototyping. For some of the biggest features that I worked on live games, we even did a version of a feature soft launch. Basically, a design of a really significantly big new feature, especially the ones that add new layers to the metagame, is equivalent to a design of a minigame!

Opportunity Cost of New Games

As game makers, we want to make games. There is a certain profound feeling of accomplishment from bringing joy to other people. Yet, making games is hard. Making good games is even harder.

There are many ways one can start a new gaming project. You can do it alone. There have been many successful games created entirely by one person. See, for example, Braid, Stardew Valley, or Cave Story (Scully, 2020). You can go about it with a certain dose of artistic naivete, conscious or not, and just start making whatever feels good to you.

However, most likely you will be starting a new project as a part of a team, with some sort of strategy aimed at minimizing the risk of failure.

I have seen and been a part of many discussions about how a team should approach starting a new game project. These discussions often touch upon a wide variety of topics, from how to evaluate the potential game ideas and marketability to development schedules to the optimal team sizes, etc. These are all valid discussions and certainly worth paying attention to. However, it is sometimes very hard to walk away from these with any actionable conclusions. Sure, there is a multitude of things to try or to avoid, but how do you know what is the right strategy for your team?

At least a part of the problem, it seems to me, comes from the fact that our industry is actually quite diverse. It is diverse, especially in terms of the types of game development teams and the circumstances in which they operate. Broadly speaking, the game development teams working on their own projects fall into three big categories:

- gaming startups fueled by investment money and looking for a secure revenue stream.
- established independent companies with a portfolio of solidly performing games.
- teams working within large gaming companies such as EA, Zynga, Activision Blizzard, Bethesda, Ubisoft, Tencent, etc.

When embarking on a new game development project, teams from these three categories face fundamentally different challenges. Surprisingly, this difference is not always entirely clear, even to people with extensive industry experience.

As always, I am writing this from the perspective of launching a game as a service, where the live service of the game is supposed to keep bringing revenue for a prolonged period of time.

Teams in startups usually do not have many people from the outside insisting on particular design solutions. They operate on tight budgets, and their survival depends upon the game hitting certain key performance indicators (KPIs), be it retention numbers to justify additional investment in user acquisition or monetization as a direct source of revenue. During the development phase, external stakeholders will focus on the team hitting the schedule, and after launching the game, they will focus on its KPIs.

A limited runway is not a problem if the airplane you are launching actually takes off into the sky. But, if your product stalls, crashes, and burns, the fire can easily swallow the whole team. Investment fuel has been spent on something that flew like a brick. Therefore, if you are part of a startup, the biggest risk you are facing is releasing a game that will not perform!

Teams working within large companies operate in a different world. Typically, such teams need to get organizational support in order to even launch their game. Organizations are not monolithic. They have a complex internal structure defined by departments with different functions and operating with different sets of goals. The team effectively needs to get buy-in from a shifting set of different operators. In general, all these people will try to minimize the risk to their own small and carefully demarcated part of the organization. Their interests often do not align with the best interests of the team and the project. This can create an environment that pushes projects into development limbo.

In this case, the task of the project leadership is not so much to deliver a great product but to convince the rest of the organization that their product is worthy of their support.

Big companies offer stability. They can take punches. Teams working from within can usually afford a mediocre launch. What they can't afford is getting stuck in production hell.

If you are a part of a big, established company, the biggest risk is that you will end up not releasing the game at all!

This is counterintuitive. Being a part of a large company should, in theory, put at your disposal enough resources, manpower, and know-how to complete your project with less risk than a small startup team can have. This difference stems from the way these two types of entities perceive the opportunity cost.

By definition, the opportunity cost is the difference in profit that would be incurred if a business entity chooses one course of action instead of another (Buchanan, 1991).

For a startup that is entirely focused on a single project, the choice is simple: either release the game that you are making or don't. The opportunity cost of not releasing the game is equal to any profit the game would make, plus the development investment that was already made. No matter how small, this profit is likely to be bigger than zero. Thus, the odds are in favor of releasing the game, no matter its quality or revenue potential.

A big multi-studio company typically has the capacity to develop several projects at the same time. It doesn't have infinite resources to invest in marketing, user acquisition, live service, etc., and it tries to allocate them in the best way. The opportunity cost calculation can easily become a comparison between a real project, imperfect as it is, and the idealized imaginary better game that they could be making.

Any concrete but imperfect project will always lose this comparison with an idealized imaginary game. Concrete projects have concrete weaknesses that can be judged, examined, and discussed. Imaginary games are, by definition utopian and perfect. They are a product of wishful thinking and usually have a form of vague ideas and lofty goals.

I have seen this happen over and over again in at least two large companies I have been working for. Projects will sometimes drag on for years before ultimately being canceled, taking with them the sunk cost of the development. Launching a game that would perform even badly would be preferable from a financial standpoint, simply because it would stand a chance of recouping just a part of the initial investment. I am not even going to mention the devastating effect that being stuck in a development limbo can have on the psyche of the creative team.

Working in an established, but independent company, seems to me, can be the sweet spot. My perception can be clouded by the fact that I have never worked for such a team. I might not know all the perils of working in such an environment.

The catch here is that the line between categories can be blurred. The company might have a solid portfolio of well-performing games, but a risky new project can bring it into the same financial danger as if it was a startup. On the other hand, a sufficiently large independent company will inevitably start behaving like a big corporation. This happens as soon as the mid-management layer becomes thick enough to obfuscate the responsibility of decision-making.

Innovation in Game Design

We live in a culture that tends to glorify the notion of an idea. From Newton and Edison, we celebrate the mavericks as the man of ideas. An apple falling from a tree, Archimedes' eureka moment in the bathtub, Mendeleev having a dream about the Periodic System; we build myths about the moments in which particular ideas were born.

The idea is almost always equated with something being novel, groundbreaking by its very definition, instantly good. This narrative is seductive, it paints a picture in which new ideas are floating through time and space, like perfect majestic snowflakes, waiting for a chance to land on an open mind. Once there, they will instantly blossom and be recognized for their value by the general public.

Anyone who has ever seen a successful project carried through to the end will vouch that having a good idea is only the start of a tormenting journey without a certain outcome.

And what about the ideas themselves? They certainly do not come in a vacuum. Nor do they have a value on their own, isolated from the context and the subjectivity of people who behold them (Byrne, 2005).

All of this applies to game design. People looking from the outside, the players, the journalists, YouTubers, and influencers of all sorts, have a tendency to praise the novelty of the ideas as a virtue of itself! They seem to be constantly surprised by the fact that established game companies seem to shun novelty and resort to the repetition of tried and true concepts.

This looks like a paradox. If the public wants unique novel ideas, why don't we just keep on making unique and novel games?

Well, the answer to this, as always, lies in human nature. People often say one thing and do something completely the opposite.

Even if you have a brilliant new game idea in your hands and your team and you manage to develop it into a form presentable to people, your idea will face the ultimate test: player expectations.

Player Expectations

Our mind does not face the world blank. Instead, we form *Mental Models*. A mental model or *Schema* is what permits us to walk into a store we have never been to before and still be able to find where the milk and cookies are. This is an elaborate set of preconceptions that our brain constructs about the world. They manifest as our expectations of how things ought to work. Our brain applies them to virtually everything in life, video games included.

Our mental models are something that we usually do not think about. Instead, we take them for granted and go about navigating our lives using them.

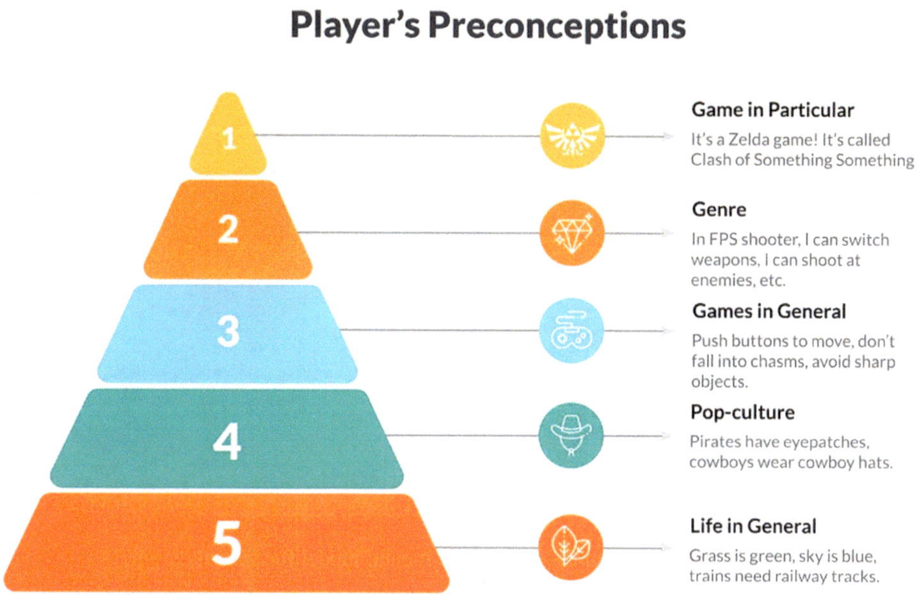

Player's Preconceptions

Game in Particular

It's a Zelda game! It's called Clash of Something Something

Genre

In FPS shooter, I can switch weapons, I can shoot at enemies, etc.

Games in General

Push buttons to move, don't fall into chasms, avoid sharp objects.

Pop-culture

Pirates have eyepatches, cowboys wear cowboy hats.

Life in General

Grass is green, sky is blue, trains need railway tracks.

Fig. 8.5 Pyramid of Preconceptions

A new player does not have any first-hand experience with your game. However, he doesn't approach your game with an empty and totally open mind. Instead, he arrives with a set of preconceptions about the game.

These preconceptions can come from a multitude of sources. In a way, you can think of them as a sort of pyramid, see Fig. 8.5.

At the bottom of the pyramid are the preconceptions that a player might carry from life and nature in general. How cars behave, how airplanes fly, how a tropical island should look, the fact plants are green by default, etc. (Fig. 8.6).

The second layer of the pyramid is players' preconceptions carried over from pop culture and other forms of entertainment. Superheroes come in all shapes and forms, but you will almost recognize one when you see one. You will instantly recognize a Western setting or Medieval European setting, although it is safe to assume you have never lived in any of these ages.

The third floor is formed by players' expectations about video games in general. We expect certain things from all video games that we come in contact with. For example, one main thing is that we expect a game to be interactive. We also usually expect to be presented with some visuals. Individual players will have expectations about the quality and style of visuals as well.

In one stage above these, we find players' preconceptions about a particular genre of the games. Most players who pick up playing a new FPS game will already be familiar,

Fig. 8.6 An imaginary alien creatures—fitting our mental model for animals

to at least some degree, with the particular tropes of the genre. They will have certain expectations about the control scheme, they will expect a choice of weapons, etc.

Finally, at the top of the pyramid are the expectations that the players will have about your game in particular. Maybe your game is part of a larger franchise that the player is familiar with. Maybe your player has seen the trailer or an ad about the game. He may have read the review in a magazine or a blog. These will all form a player's opinion about the game even before he starts it for the first time.

This is true even if the player decides to install your game just by seeing its icon in his phone's Play Store. It takes about 500 ms for a person to form an opinion about a piece of graphics. Half a second is all it takes for some sort of mental model to be formed!

Balancing Act

We validate everything that our senses register against the mental models. The value of any model is in its predictive power. If reality doesn't conform to the expectations produced by the model, the model needs to be changed.

The difference between the mental model and reality causes what is known as Cognitive Dissonance, an unpleasant feeling that we need to overcome. In the case of video games, the player will try to resolve this cognitive dissonance in any way possible. Often, the simplest way to do this is to abandon the game altogether.

Everyone doing any sort of creative work strives for originality. This is, after all, what makes a new creation so compelling. Your game, however, needs to respect the player's

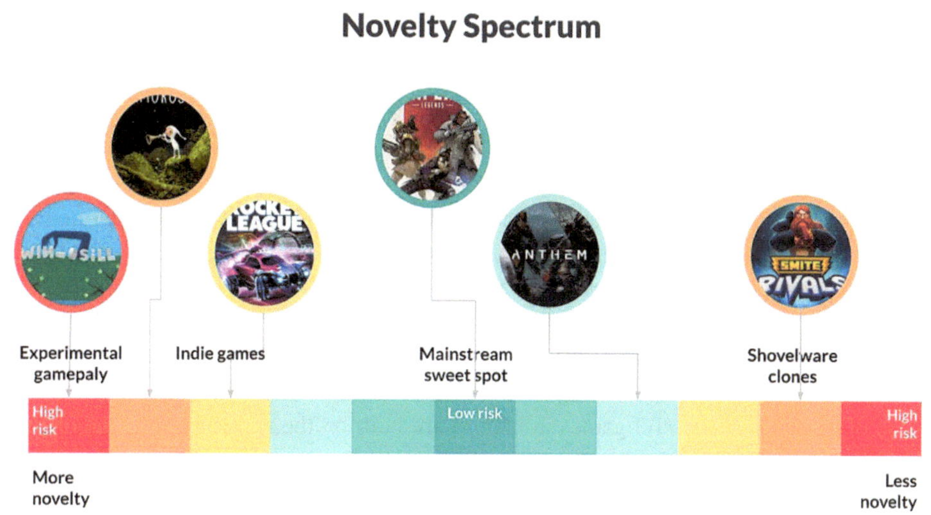

Novelty Spectrum

Fig. 8.7 Novelty spectrum

preconceptions. It needs to conform at least to a certain degree to the player's mental model.

This is a balancing act. Create something too outlandish, unlike anything that anyone has ever seen, and it will be rejected by a large majority of people. On the flip side, if you opt to create something too familiar, perhaps a carbon copy of an existing game, potential players will have very few reasons to choose your creation over the original.

Personal thresholds for novelty vary from person to person. The proportion of original ideas and well-established concepts is directly related to the amount of risk that you are willing to take. This is a balancing act that every game development team needs to perform.

Big productions require big production budgets. To justify big investments, established companies need to achieve big revenues. In order to do so, they will always try to reduce the risk.

As we have seen, the risk exists on both ends of the novelty spectrum (Fig. 8.7)!

Something Old, Something New...

To resolve this problem, the game teams will often resort to a bag of tricks that can include the following:

- *Incremental innovation*—Doom is the granddaddy for all the FPS games that exist today. However, it was not the first, it was built on top of concepts established by an even older game called Wolfenstein 3D. Most games that exist today are an evolution of some earlier concept.

- *Recombination of elements*—Elements of one type of game can be transferred into an entirely different genre. For example, Battle Pass mechanics were pioneered in the Battle Royale games and are now applied even in the city builders.
- *Simplifications*—As concepts evolve, they can get more complex. New layers are added to existing ones. This is a natural outcome of incremental innovation. Sometimes, it is good to take a step back and try to simplify existing concepts. Ideally, this can lead to a new, more refined design, one which can be appealing to an entirely new audience. Supercell's Clash Royale is a radical simplification of the Multiplayer online battle arena (MOBA) concept developed in games such as DOTA.

Some of the most successful games in history are built on the application of one or more of these strategies. The familiarity with the existing concepts ensures that the player's mental models will be respected. New elements ensure novelty. The combination of the two can result in a stroke of inspiration that happens out of the blue.

Something old, something new, something borrowed, something blue.

Evaluating Ideas

Making video games has always been my dream job. It is a dream job for so many people. Just like movies or TV, or even more, everyone has an idea for a new video game. Having ideas is great. It's fun. It's getting that idea materialized in the form of a fully-fledged piece of software that it's hard. In this sense, video games are not different than any other type of invention. Some ideas are the next cell phone, while others are the next Juicero.

There are a million reasons why you would want to make a video game. Games can be fun personal projects, or they can be works of art. Still, even the staunchest indie developers would not scoff at commercial success. It pays off to think about games in those terms.

Okay, so a video game is a product. This is an industry. The industry needs to generate revenue, at least enough of it, to cover the cost of development of the product. Gaming is also a peculiar industry, one in which revenues can reach such bewildering heights to cloud the judgment of even the soberest of minds.

In order for a product, any product, not just a game idea, to hit a home run, three things need to come together:

- the idea,
- market opportunity,
- technical feasibility.

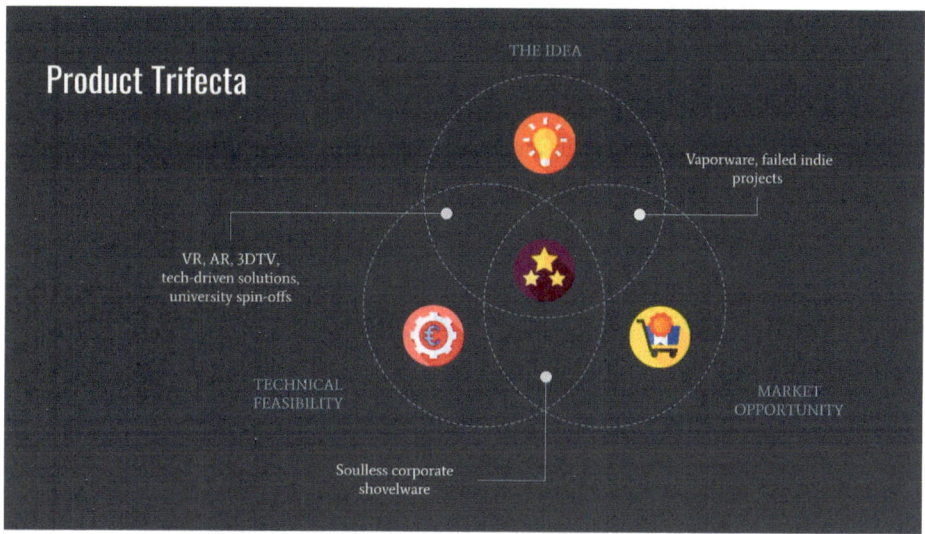

Fig. 8.8 Venn diagram of components of the product trifecta

Figure 8.8 shows the Venn diagram of these three components. It feels like I am iterating the obvious but bear with me. It is important to examine the interplay of these three things. It also pays to clarify what I mean by these three things.

The Idea

We live in a world that glorifies the very concept of idea, from Newton's apple to Archimedes and his bathtub and Tesla stroking his family cat in the dark and looking at the sparks pop from its electrified fur. We imagine ideas as some immaterial particles floating through the universe like neutrinos, waiting to land on a receptive mind, where they can blossom and transform humanity or at least some aspect of human existence.

Reality is, unfortunately, much messier. Ideas do not spark out of a vacuum. They are usually incremental steps, results of years of tinkering with the same familiar elements. They are often a product of the work of a multitude of people. They themselves rarely have any merit. In order to actually blossom, they need the two other circles of our Venn diagram.

Market Opportunity

I know that this sounds dry and materialistic, especially in contrast to the sparkling vision of the ideas themselves. However, this is the hard reality of the world we live in. The value of anything is defined only in the context of human society. Simply put, if you are making something, it should have at least someone who needs it and is willing to pay for it. Market opportunity constitutes just that, i.e., a significant number of people willing to pay money for exactly that particular product. In terms of games, this means that there are people willing to pick your game out of hundreds of thousands of new games that come to market every year.

Technical Feasibility

This last bit might seem straightforward and easy to understand, but it can actually be surprisingly multifaceted. The true magic of Apple is not that they are able to make a product as slick as an iPhone. It's that they are able to make it and sell it at a production margin that allows them to rack in enormous products.

Technical feasibility comes in two different degrees of magnitude. It can represent the ability of ANY team to execute the vision of the product. This is something that, arguably, a project of terraforming Mars would fail at the moment. Some projects are simply beyond our current technology. It doesn't mean that they will forever remain as such, but for the time being, they simply are.

It can also represent the ability of YOUR team to execute this vision. One typical mistake that novice game development teams make is to bite more than they can chew. This trap is very easy to fall into. The less experience you have, the less likely you are to be able to identify the possible problems you might encounter. Just because some other team could execute a similar vision, it doesn't mean that the team you have with the resources you have can do it also.

Technical feasibility also needs to take into account the time constraints and resources at the disposal of the team. Yes, an infinite number of monkeys, with an infinite amount of typewriters, could eventually write again all Shakespeare's works, but you do not want to be working on another piece of vaporware, Star Citizen notwithstanding.

The Intersections

It is at the intersection of these three things that we find the most successful products, the ones that have left their mark in the history of their respective fields.

Examining successful examples can provide valuable insight into how to do things the right way. After all, surely these are the blueprints to success. However, by definition, it is marred by survivorship bias. The path to success is strewn by corpses of numerous ideas, concepts, products, and enterprises that have in some way failed.

The three other smaller intersections on the Venn diagram provide some valuable insight as well.

No Idea

What happens if a clear market opportunity exists and the team has the technical capabilities to execute the product vision, but the idea is lacking? I should have colored this segment of the graph beige or cornflower blue. This is the area in which most products of corporate logic reside. In the world of gaming, this section is occupied by all the bland, derivative games, knock-offs, and shovelware. The games were produced simply because some other game made money elsewhere.

No Market

You might have a brilliant idea and the technical capability to execute it, but do you actually have the market for this thing? This is a section of the graph in which concepts such as Augmented Reality and Virtual Reality seem to be forever stuck. Yes, these things look good on paper, but the cruel hand of the market keeps crushing them in regular time intervals. No, if you build it, they will come. It does not work in real life.

I spent a considerable amount of my life in academia doing post-grad studies at a technical university. Most of the university spin-offs that I have seen have suffered from the same problem. Yes, we have this cool tech, and we have done years of research on it, so let's find an application for it. This sector is a graveyard of technology-driven solutions.

No Feasibility

The old adage of 10% of inspiration and 90% of perspiration is certainly true. Good ideas are a dime a dozen. There might even be a clear market need for the idea that you have, but this is all in vain if you do not have the capacity to execute this idea at a certain minimal level of quality.

This area of the graph is where all overambitious projects will inevitably find their resting place. In the world of gaming, this is the death coast peppered with the shipwrecks of various indie games, my own amateur projects, and failed Kickstarter. This is where the proverbial vaporware lurks.

If you have an idea that is solid and you are able to identify the marketing niche for it but do not have the capabilities to actually make it, this means it is a good idea for some other team.

All of my own solo game development projects firmly fall into this category.

Way Forward

If you are at the start of a new project and you are examining a concept, try to determine as soon as possible and as objectively as possible where in this diagram it falls. If it is missing two out of three elements, it is safe to assume that it should be discarded. If, however, you have hit two out of three targets, try to think of a way to get hold of the remaining missing element.

If the idea is lacking, do brainstorming and try to think outside of the box about how to bring something new to the concept.

If you lack the technical capabilities to execute the vision, changing it and cutting the scope might be the answer. Also, hiring people and acquiring the needed tech might help.

Dealing with the lack of market need is arguably the hardest to navigate. It is probably a futile effort.

Product Pipeline

The process of bringing a new game to the market is often referred to as product pipeline. Starting a new game studio is somewhat akin to starting a new bend. The chemistry of the initial team members can be a more important factor than their actual abilities and skills. New studios are often driven by vision and pure desire to create new games and simply bring something new into the world. Releasing one game can be hard. Releasing more than one successful project is the order of magnitude harder. Having the ability to consistently release new games with any degree of success is an entirely different ball game.

Building an efficient product pipeline able to deliver new titles in regular time intervals is what the middle management layer of any established game company is spending the bulk of their working time. This is also where things tend to become quite murky. Working at four established gaming companies, I alone have seen at least a dozen different product pipeline designs. Anyone who has spent any time in the industry has, without a doubt, encountered the discussions around the product pipeline.

The purpose of an efficient product pipeline is to minimize the risks involved in making something new and creative and maximize the chances of its success. As someone said, there is nothing more tragic than a capable, team executing a flawed plan. As we mentioned earlier there is always a certain opportunity cost associated with starting a new game project. A well-designed product pipeline should in theory ensure that the tram invests its efforts into the project with the smallest associated opportunity cost. This is of course, much easier in theory than it is in practice.

Broadly speaking all product pipelines fall into three distinct categories. Each of these categories is based on one of the three key components of successful product design that we mentioned in the previous section: the idea, the technology, and the market opportunity.

Idea-Driven Pipeline

One of the quirks of the game industry is that everyone has ideas for new games. The urge to bring your own game ideas to life is why most people become game designers. It is definitely true in my case. It is also true for a large number of people working in other roles within the industry, artists, programmers, producers, etc.

Therefore, an initial idea for a new game can come from anyone within the game development team. It is safe to say that the games industry will never run out of fresh new ideas. Which of those ideas can be developed into a successful product, i.e. one that millions of people will play, find joy in, and remember for the years to come, is a billion-dollar question. Quite literally.

Any gaming studio driven by a strong design vision will tend to gravitate towards an idea-driven type of product pipeline. This is especially true for a lot of gaming startups and indie studios but also for studios within established companies. Indeed some of the most successful gaming companies such as Nintendo, and Supercell will organize their pipeline in this way.

The process of developing games in this framework begins with the idea. The rough idea is obviously just the nucleus. Quite often, the idea itself cannot even be evaluated in the proper sense. It's market fit and technical feasibility cannot be estimated directly. Simply put there are too many unknowns and too many possibilities packed inside any idea to make this very difficult. The metaphor of a subatomic particle for the idea is a very adequate one. Even the simplest idea can contain so much unexplored potential that it could explode into something big or more likely fizzle down and disappear.

In order to reduce the uncertainty inherent to the rough idea, we need to refine it. This is by definition the second phase of the process. The idea becomes something much more tangible, something that can be examined and critically evaluated. It becomes a concept. The job of transforming an idea into a concept of the game is usually the job of the design team.

The concept is again a very nebulous term. It could mean various things to various people. Indeed it can mean various things in various contexts. In general, a game concept should contain enough detail to permit the next two steps in the pipeline to take place. These two steps are market research and assessment of the technological feasibility of the development of the project. There is no fixed order in which these two steps need to be taken. Since they usually involve people with quite different areas of expertise they are quite often done in parallel.

Market research aims to determine the marketing fit of the idea. It tries to answer some very straightforward but very important questions:

- Are there enough people who would like to play such a game? Meaning, does this game have a large enough potential target audience?
- Are these people already playing a similar game? Meaning, who are we going to compete with?

Furthermore, market research tries to determine the potential revenue that the game can be expected to make. Finally, this part of the process also tries to predict the potential cost of marketing of such a game.

The technical feasibility step of the process tries to determine if the team has the ability to carry out the project until a successful conclusion. It is trying to foresee the technical requirements of building a proposed game, to assess the needs in terms of tools, infrastructure, and skills of the team members. Finally, it tries to determine the potential cost of the development of such a game.

The technical feasibility step of the process tries to answer the following questions:

- What is the reasonable scope of the product vision?
- Do we have the right technology needed to develop this vision?
- If not, what other technology do we need to develop or obtain?
- Do we have the right people with the needed skills to execute this project?
- What tools do these people need?
- Can we do this in a reasonable amount of time?
- How much will all this going to cost?

The cost of development of the game is added to the cost of marketing the game and compared against the potential revenue estimate. This is how the potential Return on Investment of a project is determined (ROI). Figure 8.9 represents a diagram of an idea-driven product pipeline.

Opportunity-Driven Pipeline

As I mentioned before, the game industry will never run out of ideas. Picking the right idea to develop and implement is the hard part of the process. Many game companies, especially large and established ones tend to take this for granted. They treat ideas as a commodity, counting on their ability to attract and hire enough creative people capable

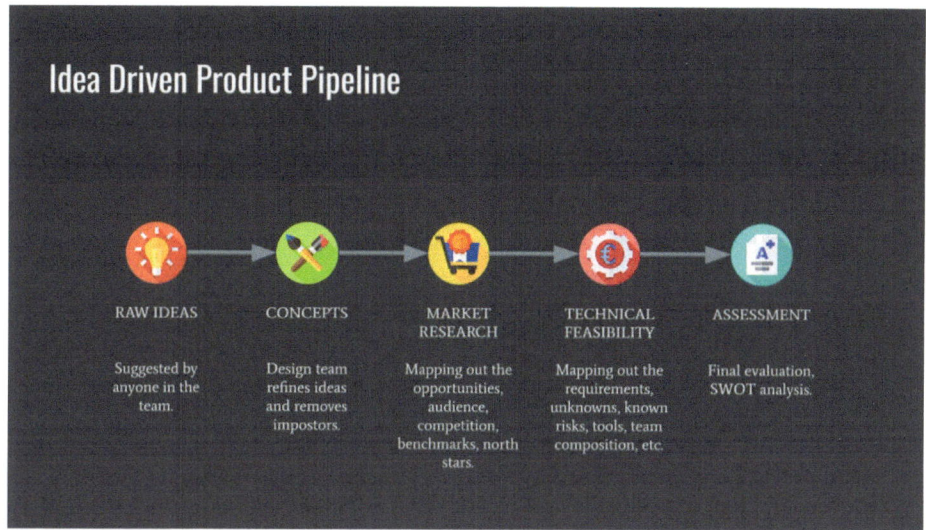

Fig. 8.9 Idea driven product pipeline

of producing good ideas on demand. For a similar reason such companies also tend to codify the technology. They typically have the resources needed to execute even the most ambitious projects.

These companies organize their pipelines around market research. In this context, the answers that the market research tries to answer are slightly different. The main questions are:

- Is there a segment of the market unaddressed by existing games? Meaning is there something that competition is overlooking? Some section of the audience not served by the competition?
- Is this niche big enough to be worthwhile?

The next step of this type of process revolves around the idea again. However, the idea in this case needs to fit into the market hole identified by the market research. This already puts much harder constraints on the type of ideas that can be considered. If the market research has determined that there is a big enough niche of potential players interested in an FPS game set in paleolithic Siberia then we are not even going to discuss making a real-time strategy or a match-3 game.

Obviously, the technical feasibility of the project needs to be determined, and to do so we need to have at least a concept of the game. This determines the order of the remaining steps of the process. A diagram of such a pipeline is shown in Fig. 8.10.

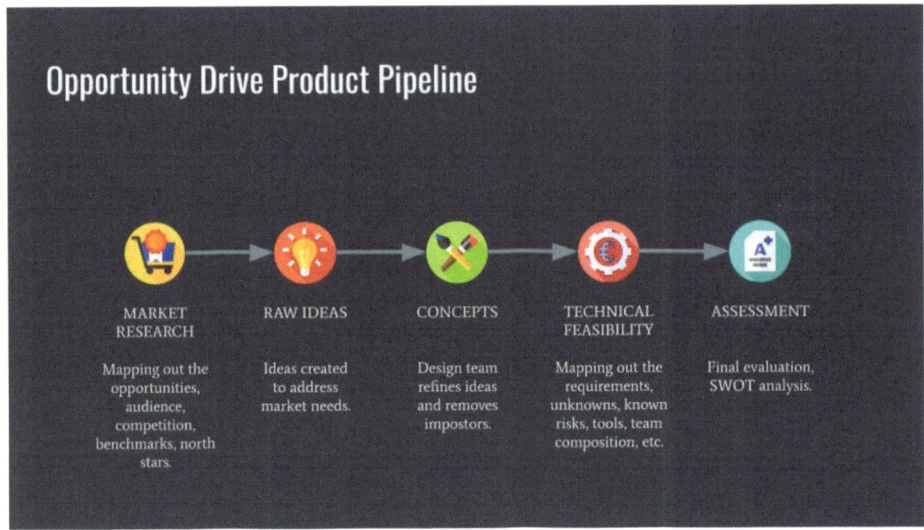

Fig. 8.10 Opportunity driven product pipeline

Technology-Driven Pipelin

Finally, one other type of pipeline exists. This is a technology-driven pipeline. Technology in this case can mean the actual software or hardware tech, for example, NaturalMotion Limited now owned by Zynga started off as a company specializing in real-time animation software and started developing games to showcase their technology. Technology can also mean the accumulated know-how and tools that the team has accumulated over the years. A team having expertise in developing casual free-to-play match-3 games will tend to focus on similar types of games and be careful before venturing into building a battle royale FPS.

The next step of this process will again involve coming up with the idea and developing the idea sufficiently in order to asses its market potential. Figure 8.11 is a diagram of such a pipeline.

Selling Your Ideas

If there is one thing that I wish I knew as a high school kid, it would be the following. Knowing how to sell your ideas is one of the most important skills in life. By "selling," I mean being able to convince others that things you have in your mind are worth their effort or at least attention. In fact, if I think about it. This is what I have been doing most of my life. Passing exams in school or at the university was all about convincing the professors

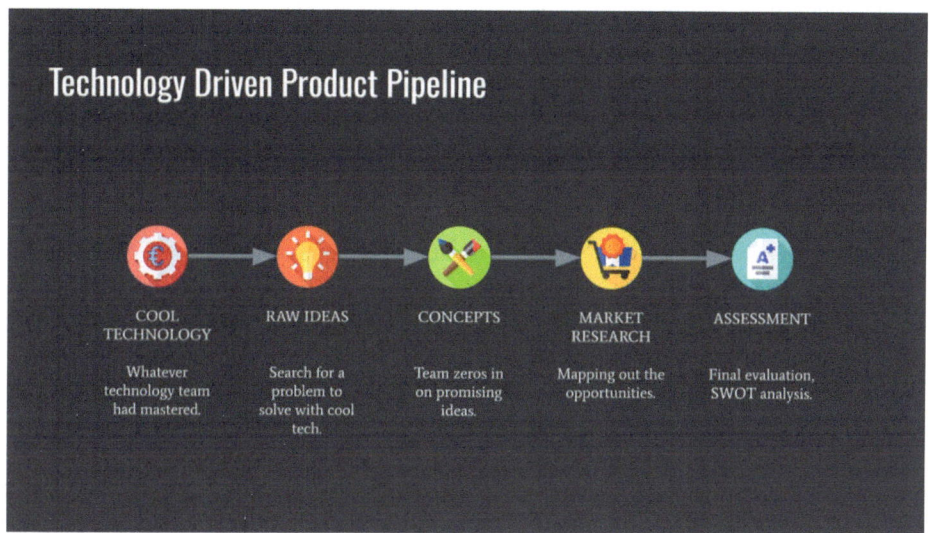

Fig. 8.11 Technology driven product pipeline

that the ideas and notions I had in my head were aligned with their expectations. Later, working at the university as a post-graduate researcher, I was expected to produce papers, i.e., scientific publications on topics of my research. That too, was all about selling my ideas, i.e., convincing peer reviewers that my results were novel enough and interesting enough to be worth publishing.

As a game designer, I face this task over and over, each time I need to present a proposal for a new feature, let alone a whole new game. Knowing how to sell your ideas is an essential skill for anyone working in a creative team. It is also something I feel I will continue learning and improving for the rest of my life. Here are a couple of things I learned along the way.

First of all, when discussing ideas, learn to be dispassionate. We tend to get attached to our ideas. We tend to treat them as our babies. This, in turn, can lead to a lot of emotional distress when our ideas get dissected by others or, even worse, rejected. This is the wrong way of seeing things. It doesn't matter whose idea it is as long as it improves the project you are working on. Furthermore, you are almost never going to be evaluated on the basis of the quality of your ideas but on execution.

Next thing, know what you actually want to present. Work on your idea until it is clear in your head. Develop it into a concept with enough details to provide relevant information to your listeners.

Know your audience. This is actually crucial. As a designer, you are most likely going to need to present your idea to several very different groups of people. Most likely, you are going to need some sort of approval from other key stakeholders. These may well be people higher than you in the team or company hierarchy. They are most likely going to want to understand how your proposal aligns with the big picture, i.e., how it supports the high-level goals of the project. On the other hand, the rest of the team, the programmers, the artists, the QA, etc., will want to know about your idea on a more detailed level. They will be interested in the technical details of your proposal. People in charge of production scheduling, such as producers, development managers, and development directors, will be interested in the scope of your idea, i.e., the scope of the development effort needed to realize it. Other disciplines will have their own specific questions.

Do not try to make a one-size-fits-all presentation. This is a bad approach. It will inevitably burden people from each of these disparate groups with a ton of detail that they do not particularly care about. Even worse, presenting needless details to a group interested primarily in the high-level vision of the proposal might lead to a situation where someone latches onto a particular detail, which might be inconsequential in the long run and derail the whole conversation. I have learned this mistake the hard way.

In order to systematize things in your own mind, you might want to create one big presentation that contains slides describing various aspects of your proposal. Use it as a basis, and extract just the needed slides into separate presentations intended for the particular discussions with the particular audiences. This is the method that I often use.

Start from the general and work your way into the details. Always begin with a set of goals. Great writer Kurt Vonnegut had simple advice about making a novel. Each scene in the novel should either move the plot forward or reveal something about the characters. Otherwise, it has no place in the book. This applies to your idea. If you can summarise its reason for existence in a set of three to five goals, it probably has no place in your project. Vonnegut's advice applies to the structure of each of your presentations. Each slide has to have a purpose.

After stating the goals, when proposing a new feature for an existing game, be it a live one or one that is being developed, I sometimes list the stuff that already exists in the game that I am building upon. This is not always the case.

Proceed with describing a high-level concept. This should neatly fit onto one single slide. Use the remaining slides to elaborate on the details.

Keep the text minimal. Iggy Pop has great advice about writing lyrics for rock songs. Twenty-five words or less! Keep the text on your slides in three to five bullet points. If you can't compress what you want to say into three to five short lines, it is a good signal that the content should be divided into several slides.

Do not worry if you are not including all the details in your slides, even in a deck intended for the people who will be doing the implementation. Your PowerPoints should not be your design documentation. Design documents should reside elsewhere and be maintained differently. They should include all relevant details.

Use pictures, diagrams, and illustrations as much as possible. A picture is worth a thousand words. This is absolutely true. Never put more than two or three pictures that are meant to illustrate different things on the same slide.

These short, focused presentations intended for use in a particular meeting with a particular group of listeners in mind should ideally be between five to seven slides long. This doesn't include the title slide and the ending slide. The last slide should give a clear signal that this is the end of your presentation. If you can't fit everything into a maximum of 10 slides, your presentation is probably not focused enough. Reconsider its content.

Sometimes, this is not possible. You might want to include slides with additional details. Include them in the clearly marked appendix of the presentation.

You do not need to have all the details already worked out before presenting the idea. Be clear about the unknowns. Be clear about open design questions. For some of them, you might not have the answer right away, but you could be confident that you would find one in due time. For example, you do not need to know the exact number of reward tiers that a Premium Pass would have in the end, just a ballpark estimate, and this number needs to be configurable. On the other hand, some open design questions will require more work, discussions, and experimentation. Be open about this. Your team is there to help you find solutions.

Do not try to think outside of your domain, especially when talking to the other disciplines in the team. Your team is full of experts in their own domains. Do not try to think for them. Ask them for their input. Above all ask them for input and feedback on their area of expertise, but ask them also for general feedback on the idea.

Finally, keep in mind the purpose of each meeting. Some meetings are about getting approval to proceed or not with a certain design. They should end in one of the three clear outcomes: either everyone agrees that the design can proceed forward, or the design should be totally abandoned, or the design should be significantly reworked. In this last case, do not try to solve design problems in the same meeting. Gather feedback and iterate on the design, alone or with a proper group of people. Some meetings are going to be kickoff meetings in which everyone needs to get acquainted with the design to some level of detail. These meetings should end with everyone aligned on the main points of the design. The details should be written in the design documentation. Everyone should leave the meeting with the knowledge about where to find those details.

Iterative Game Design

I have mentioned several times already the importance of iterations in a game development process. Any design process, and indeed any creative work, is not simply a journey from an idea to a finished work of art. It is a complicated search for the solution that oftentimes takes you down the wrong path. The creative process is iterative by nature. You begin with an idea and create a vision. You try to get as close as possible to that vision from the point in which you happen to be at the moment. You produce something, but eventually, you need to step back and take a look to see how far you are from your vision, lest you lose your way on this journey.

Design is about finding solutions to practical problems., but finding an optimal solution is not easy. Sometimes, there is no optimal solution. Sometimes, it is a matter of choosing between several equally suboptimal options. In any case, to maintain your progress, you need to receive feedback and input from the outside that will help you on your journey. The faster you iterate on your ideas, the more ground you will able to cover, getting closer to your goal.

The iteration cycle is simple enough. Create a design, implement it, gather some feedback, make changes to the design, and repeat the process. Figure 8.12 shows a schematic diagram of this cycle.

This simple structure permeates the whole game development process throughout all of its phases. Only the level of implementation of design gradually changes and becomes more intricate. In the beginning, the design is just in the head of the designer or scribbled on a whiteboard. The "implementation", its embodiment, in this case, is elaborated on a piece of paper or a deck of slides. Feedback takes the form of a discussion. Feedback is always a discussion. Later on, design resides in some form of design documentation. The implementation could be a piece of code or a playable prototype, but the cycle remains the same. Come up with an idea, implement it somehow, and test it out. Se what works and what doesn't and make changes accordingly.

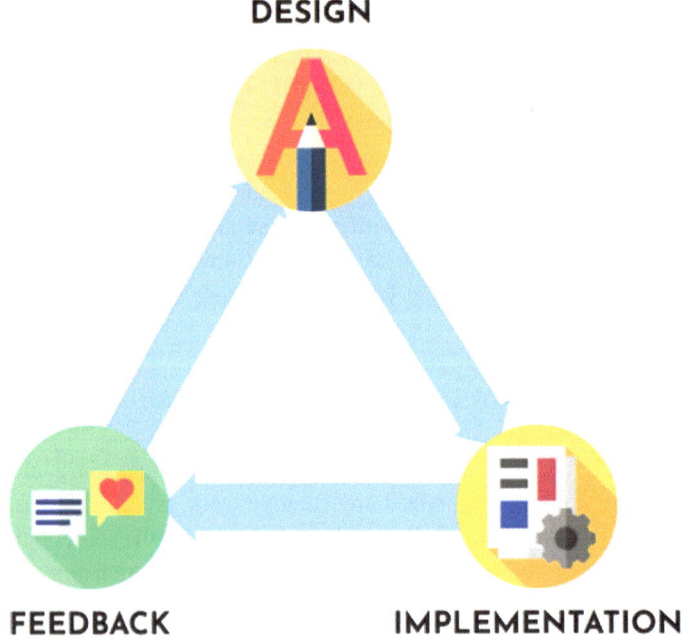

Fig. 8.12 Diagram of iterative design

These feedback sessions can also take various forms, from ad hoc meetings of team members to internal playtests and focus group testing to limited market tests and soft launches. Even when a new game or a feature goes live in front of a global audience, this too, is a chance to gather valuable feedback that you can and should use to improve your designs.

One thing that all of these feedback sessions have in common is that they should include at least some people who were not involved with the design process and implementation process itself. Seeing the work with a fresh set of eyes is invaluable. You can create design specifications in a small group or even alone. Having too many people involved in this part of the process is usually counterproductive. Expand this group to learn about your design, its qualities, and shortcomings.

During the feedback sessions, focus on gathering feedback. Do not try to solve the design problems then and there. Collect the feedback and process it afterward. Filter it into two large groups: actionable and nonactionable feedback. Actionable feedback is something that you should act upon. It needs to be relevant, i.e., something that is fundamental to the success of the design or something that is expected to have some material impact or bring some new quality to the design. It is also something that you are able to act upon, i.e., something doable in a reasonable amount of time with the available resources. Everything else is nonactionable. This doesn't imply that nonactionable feedback should be forever ignored. Some things that lend in this plie simply need to be postponed due to practical reasons. Your server programmer might be on holiday, or new art assets are expected to arrive next week.

Prioritize actionable feedback according to two criteria: expected impact and difficulty of implementation. These two variables allow us to divide all action points into four quadrants, as shown in Fig. 8.13.

Quadrant A is populated by the items with potentially high impact but relatively low implementation cost. These are the low-hanging fruits that should be on the top of the priority list. In contrast to these, quadrant D is populated with items that should be ignored, i.e., low-impact items that require high implementation effort. The items in quadrants B and C are the ones that require more consideration. Quadrant B is populated with the hard nuts to crack, things that require a significant effort but are also expected to produce significant improvements. Tackle these after you have exhausted all the items in quadrant A. Items that require low effort but are expected to yield only small changes in the design are nice to have items and should be tackled only if all the major problems have been solved.

When making changes to the design, try to limit the number of changes. Try to implement changes that are as disjoint as possible. Too many changes that affect the same aspects of the design make it hard to evaluate the results. You simply cannot know if the new results are a product of change A, change B, or both. What if change A and change B actually create opposite effects, canceling each other partially? Keep things as separate as possible. All changes that you make need to be there for a reason. Don't make changes

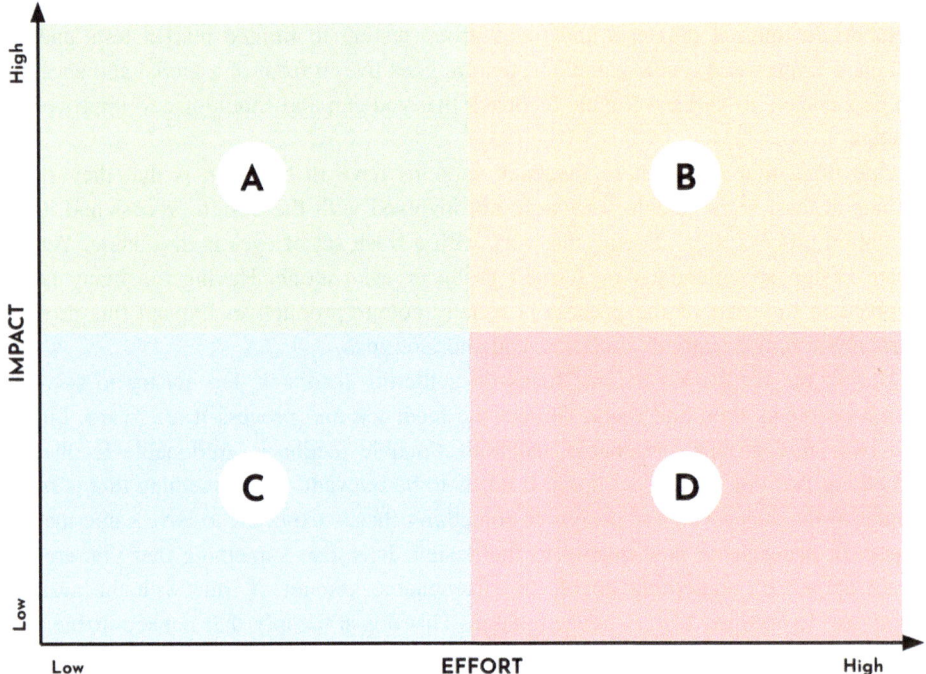

Fig. 8.13 Priority quadrants

for the sake of making changes. Each change has the potential to make things worse. Don't fix it if it isn't broken.

If you are unsure how to tackle some of the important feedback points, i.e., if you are facing a design problem, use separate problem-solving meetings to arrive at a set of possible solutions, but keep problem-solving meetings and feedback sessions separate. I know that this is often hard to maintain. People always seem to forget this, and I myself have stumbled on that rock more times than I can count, but mixing the two usually just creates a mess. Feedback meetings should be about identifying the problems. Problem-solving meetings should be about finding solutions.

Finally, know when to stop. In my experience, this seems to be the hardest thing to master. The iterative process can't go to infinity. At some point, changes that you make will stop yielding improvements, and you will start to run into circles. The distinction between what is a change and what is an improvement is subjective and can depend on the context. However, if you catch yourself constantly flip-flopping between two possible alternatives, this is a dead giveaway that it is time to stop. At this point, you are better off rejecting both alternatives and trying to figure out an entirely different approach. Alternatively, if this is not possible, just pick either of the alternatives and roll with it without looking back. Moving even in the wrong direction is always preferable to being forever stuck in a dead loop. This way, you will at least produce something.

GROW Model

After so many years of working as a game designer on a live title and after numerous updates, there is one learning I believe I share with everyone who has been in a similar position: game design is a messy process.

Maybe you start off with an idea for a cool game feature, maybe you start off with a gaping hole in one of your metrics, or maybe you start off with a vague set of goals... You brainstorm, you toss and turn the original idea, you add to it, and you subtract. You iterate, and you negotiate with your developers and artists. More often than not, you end up running in circles. More often than not, the end result turns out to be a horse designed by a committee or, even worse, a camel designed by a committee.

Anything that can streamline this process that can make it less convoluted is always welcome. In talking to my coach, I came across something that I found interesting and applicable. The GROW model has been a part of the corporate coaching toolbox for about three decades since it was published by John Whitmore in his book Coaching for Performance (Whitmore, 2017) and Max Landsberg in The Tao of Coaching (Landsberg, 2015).

The GROW model is a goal-setting and problem-solving method that consists of four distinct steps, which produce the letters of its acronym:

- G stands for Goals,
- R stands for Reality,
- stands for Options,
- W stands, well, for "will" or for steps that one decides to take.

It is an intuitive yet versatile framework that can be used for a wide variety of practical problems. In what follows, I'll elaborate on my take on how this method can be applied to game design tasks.

G is for Goals

A video game is not a monolithic piece of software. Neither is a single chunk of design. Rather, it is an intricate system composed of a series of interconnected features. Here a feature can be anything, from the control scheme or a gameplay mechanic to a bit of UI.

Let's assume that you want to add a feature to your game. I am speaking from the experience of creating features for an already published title that we maintain and develop as a live service. Now is a good time to remember the writing advice formulated by Kurt Vonnegut: "Every sentence must do one of two things—reveal character or advance the action."

Every feature in the game needs to have a purpose for its existence. Every feature needs to fit in with the rest of the big picture. Features need to communicate with other features. They are interconnected in a myriad of ways. New features are often built on top of other already existing ones.

Therefore, your design process, by necessity, begins with a set of goals. At first glance, this may sound obvious, yet in practice, it is often overlooked. I have seen teams get into a design process with a very ill-defined loose set of goals that can even be contradictory, or even worse, jump straight from an idea to intricate the design without giving much thought about the purpose of the thing that they are designing. This is the first and biggest pitfall! Avoid it at all costs!

Spend time to define a set of goals that your design needs to fulfill! This can in itself be messy, but spending time at this part of the process will save you an immense amount of wasted effort later down the line! Your design can try to hit several goals. However, it needs to have at least one. Goes without saying that one should be mindful of focus and avoid trying to impose too many goals on a single design.

The very act of spelling out and writing down a set of goals is the first step in bringing clarity to the design process. There are many ways one can go about accomplishing this task, but there is one particular method that is worth looking into. If you have any experience working in a big company, chances are you have encountered the notion of SMART goals as a part of your yearly evaluation process (Doran, 1981). I know that this acronym can bring a sense of cold dread, but bear with me.

The letters in this acronym stand for the following:

- S—specific,
- M—measurable,
- A—attainable,
- R—relevant,
- T—timeboxed.

These categories can be directly applied to goals suitable for the design of a game feature.

Specific—your feature needs to have a specific purpose. It can be adding new functionality, creating a particular kind of user experience, modifying player behavior in some way, etc.

Measurable—you need to be able to define at least one quantitative metric by which you are going to evaluate the performance of your design. This can be any of the usual KPIs (Key Performance Indicators), such as engagement (a percentage of eligible players that chose to interact with the feature), day two, day seven, or day thirty retention, same monetization metric, or anything else that is applicable.

Attainable—in a practical sense, the goal needs to be attainable with the resources you have in the time frame that is at your disposal.

Relevant—no feature exists in a vacuum. Every feature is a piece of a bigger system. The goals that it needs to satisfy must be aligned with the goals of the larger product. You can specify a valid goal that is simply irrelevant to the overall structure that you are building. This is something to be avoided.

Timeboxed—time is one of the most common development constraints. It is also one of the most useful constraints that forces the team to stay focused and keep the scope manageable. In practice, the schedule for designing and developing any individual feature usually needs to align with an overarching development schedule, for example, a release cycle or development milestone deadlines.

If, after applying this methodology, you end up with a list consisting of a handful of goals that satisfy these criteria, you are on a good footing to take the next step of the process.

R is for Reality

If the previous step, the goal setting, was about deciding where you want to go, this step is all about determining the actual starting point of your journey. This in itself is a multifaceted phenomenon.

As we already noticed, no feature exists on its own. Most likely it is built on top of existing features and needs to fit in with numerous others. It is likely that it will be used as a platform for further development. Map out these dependencies the best you can. Even though some connections between features might not be instantly apparent, creating a canvas of obvious ones should help untangle the hidden ones.

This step of the process will, by necessity, involve taking stock of the current state of the project. It will also likely include examining the state of the infrastructure that the future development will make use of.

The design process is often iterative in nature. It is unlikely that the design that you are working on is already fully formed at this stage. List the open design questions and, if possible, prioritize them according to the expected impact on the rest of the designs.

Take stock of available resources, including the technical resources but also the skills of the team members. Make an inventory of tools at your disposal.

All this will help you get an understanding of what are the missing elements, what are the pieces of the feature that yet need to be created, what are potential design problems that need to be solved, and what are the tools and resources that you might require to accomplish your goal. You should be ready to take the next step.

O is for Options

After the previous two steps of the GROW method, you should have an understanding of the goal of your development journey and of your starting point, the proverbial points A and B.

In terms of the roadmap metaphor, the third step of the process is about getting a clear picture of various paths that can take you from point A to B, or at least figuring out what possible next steps you can take to get you closer to your goal.

Sometimes, these paths can be obvious. Sometimes, you are dealing with known quantities, and the road ahead is clear. In this case, the rest of the process is all about execution.

In other cases, your task at this stage will be to actually come up with a set of viable options. The corporate coaching and performance improvement literature has a considerable body of work on various methods that might be used. Many game design teams actually make intuitive use of some of them. I will list just a few.

Five Options Approach—is what you should be doing if you already have lots of possible ideas floating around. It is supposed to bring order into the chaos. Write down, list, and enumerate ideas. List the pros and cons of each one. Prioritize them according to some criteria, such as the difficulty of implementation, chance of success, or expected impact. Perhaps, apply some other well-known formalism to them, such as SWOT analysis (Ansoff, 1980).

Brainstorming Approach—is the polar opposite of the previous one. It is most useful when you have only a high-level understanding of the goal of the design but lack a clear set of ideas. There are many methods proposed elsewhere to efficiently run a brainstorming session in order to produce high-quality ideas. However, this discussion is beyond the scope of this text. Keep in mind, though, that the result of this brainstorming will most often need to be pruned in some way.

Obstacle Approach—this is applicable in the case when you encounter an obvious roadblock on your journey and you don't have a clear idea of how to remove it. The key idea of this approach is to imagine the roadblock magically removed. Try describing the reality without the big boulder in the middle of the road. What does it look like now? Ideally, this mental trick would allow you to move beyond the point at which you were stuck. By shifting the focus, you might be able to figure out a way to circumvent the problem in the first place, or at least it would allow you to apply your resources to something else potentially worthwhile.

Ideal Future Approach—involves a mental trick similar to the one that we just described, yet targeted at an even more distant point on the path. Try to imagine what the ideal outcome of the process should look like. What the reality of your project would look if you manage to achieve your design goals? Ideally, this would allow you to work your way backward, mapping out the path from the end to the place where you are standing now. This is a formalism that is akin to the way labyrinth puzzles in kids' puzzle books are solved.

Transformative Approach—is applicable when you have something that already exists that can be used as a starting point for further development. Perhaps you already have a feature that is addressing similar problems as the one you are trying to build. Perhaps all that is needed is to boost its performance or to modify its functionality in some way to make it perform on another level. This is surprisingly often the case when dealing with established projects and live services. However, this is an equally valid approach to salvaging elements of canceled projects. Before building something from scratch, it is often worth looking into what already exists that can be repurposed.

W is for Will

The result of the previous step of the process should be a list of options that you might take. In contrast, the final step of the process is about actually choosing the option that you are going to take.

Your task, therefore, is to make an informed choice, ideally the best possible one, based on your understanding of the Goals, Reality, and Options.

This choice can sometimes be hard, especially if you have several concurrent options that are competing for the same pool of resources. You might want to apply some prioritization methods using some formalism designed for this purpose. For example, you might associate two numerical values to each of the possible options, one indicating the expected impact and the other difficulty of implementation. Plotting these value pairs on a 2D diagram can help you get a clear picture of which steps you should take next.

This formal approach is reductionist by definition. Just two variables might not be enough to sufficiently describe your decision landscape. You might want to consider the alternative values, such as probability of success or risk. This process is also dependent on the precision with which you are able to determine the quantitative values you are associating with your options. This can be as much a result of guesswork as it is an exact science. To avoid at least some of these problems, use the case-by-case approach.

Ultimately, when the choice is made, the remaining part of the GROW method is equivalent to forming an action plan. In most design teams, this step would tap into the normal process of roadmap creation, sprint planning, and task breakdown.

The goal of the method that I just described is, above all, to bring some structure into an otherwise messy process of game feature design. This is not a magical solution, it is not a silver bullet that will automatically solve all your design problems. It is not even meant to be. In practice, it might not be applicable in all of its aspects, yet hopefully, it might be a useful way of looking at things.

I need to make one additional disclaimer. I am not a big fan of dogma and the dogmatic application of rigid frameworks. Every method needs to be adapted to the reality of each individual design team and a concrete game design process.

Dealing with Documentation

Monolithic design documentation is a thing of the past. At least, that is my personal experience working on mobile Free-to-Play games. When I hear the stories of the older generation of game designers about fabled Game Design Documents (GDDs) of yore, I feel something resembling envy. I have never worked with one, and I have to admit I am still amazed that people are able to produce games in this way.

Games run as a live service and evolve during their lifecycle. The methodology of their development is much more agile nowadays. The documentation needs to reflect these changes. I might not have seen a real GDD in which a game designer specifies every aspect of the game, from the control scheme to the sound effects, but I have seen all sorts of approaches when it comes to the design documentation. Different teams approach documentation in different ways. Some very successful live games are run with totally chaotic documentation consisting of a bunch of Excel sheets, PowerPoints, and random documents catered over multiple local or shared hard drives. On the other hand, you can have meticulously maintained documentation and still produce a crappy game.

Stone Librande, a game design veteran with experience working on games such as Diablo III and SimCity 2013, advocates something called One-Page Design (Librande, 2010). This approach shares quite a lot of similarities with the approach to making presentations that we discussed earlier in this chapter. It basically suggests creating a game design that can fit on one sheet of paper. Since obviously a complete description of even a moderately sized feature, let alone the whole game,m cannot fit completely on one single page, Stone suggests creating multiple such pages with different purposes in mind. One such one-page design would be created for the executives, providing them with a high-level vision of the game. Another such page would be created to outline the backend architecture of the project. Yet a different one would fit a design for a single level, and so on.

This approach highlights two important aspects of the documentation. The documentation needs to be concise, thus the space limitation. However, it needs to provide different information to different people. One of the key purposes of the documentation is to help the team align around the vision and the work that needs to be done. To achieve this goal, documentation needs to be read by the people. To do so, people need to first know where the relevant documentation can be found, and second, they need to be inclined to read it. People, in general, do not like reading, especially technical documentation, especially if it is full of details that do not concern them. Librande's one-page document solves this question as everyone is served only the most relevant information.

The disadvantage of Librande's approach is that you eventually end up with a bunch of papers. Keeping track of them might be a problem. On the other hand, it is often useful to still keep your documentation in one central place and use it as a central reference point.

Running a multiyear live service adds an additional layer of problems. As I mentioned game evolves, its features evolve with it and get changed, and the content gets added. How do you organize your documentation to reflect this? Unfortunately, I do not know a good answer, but rather a couple of different ones each with its own disadvantages.

You can organize your documentation around features, with one document dedicated to each feature, but you still need to keep track of the evolution of individual features. You can always update these documents to reflect the current state of the game, but you lose the information about its history. Sometimes, you need to go back, revert changes, or even just find out how stuff used to work. This approach also makes it hard to follow which features changed and which didn't or when was each bit of content added to the game. QA would have a problem figuring out what needs testing in each new release cycle and what can be safely skipped.

The other approach that you can take is to organize your documentation around your update cycle. Each update can have its own section containing the documents describing only the new features and content and changes made to the existing stuff. However, this approach is also not perfect. Not all aspects of the game get touched in each update. Did we change how Seasons work in U23 or U25? Were the castles introduced in U16? Digging up changes requires you to sometimes search through multiple sections for different updates.

Things get easier on the level of individual features or content packages. As we mentioned earlier, different groups need different levels of information. For example, the QA team uses documentation as a reference about how the game is expected to behave, i.e., the specification against which they are testing the implementation. They will be interested in the details of UI flow. On the other hand, developers use it as a blueprint for what they are supposed to implement. Community Managers would be interested in a totally different level of information. They are curious to know how the new feature or the content is going to be presented to the player, and so on.

To solve this problem, I take an approach similar to the one I mentioned when discussing making slide decks. I always start each individual feature page by stating the goals. Why are we making this thing, and what do we expect to achieve with it?

In the second section, I present the concept of the feature a rough outline of its basic high-level design. This serves both as an introduction to everyone and as a quick summary to people who need just this: a brief description of the feature.

The next part contains the actual functional design of the feature, including the currently up-to-date values of any variables that need to be balanced. I make sure to indicate the stuff that is still a work in progress or even just placeholders.

In addition to this, one whole subsection is dedicated to the UX, or more precisely, to the detailed UI flow in case the feature contains any player-facing UI elements. These two last sections can freely change places. I am an engineer at heart, but for many people, seeing the images of UI flow first makes understanding the operation of the feature easier.

Any additional documents, such as spreadsheets, slide decks, PDFs, etc., can be attached to this page. This is, of course, only one approach to maintaining the game documentation. It is not perfect, but it served me and my team well for multiple years. I am not suggesting that you should use this as a template for your own purposes. Rather, I am sharing this as an example.

Understanding Roadmap

Each craft has its own jargon, and so does the project management. Barbers will talk about *Buzzcuts* and *Crewcuts*, Minecraft players will talk about *godbridging*, and anyone involved with planning any project more complex than putting a frozen pizza into the microwave oven will at least once a week encounter the phrase from the title of this text.

Corporate speak is notorious (Preply, n.d.). People cringe at it for a reason. Yet as our society continues to grow in complexity and office jobs attract ever more people, the phrases that originated in the corporate environment have started to seep into every nook and cranny of mundane civilian life. So much so that we now tend to take them for granted. They have become figures of speech that we utter without thinking, taking into account only their acquired figurative meaning.

It pays off to sometimes stop, take a step back, take another look at the origins of a certain phrase, and examine its actual meaning.

Roadmap is one such metaphor (Phaal et al., 2004). People tend to use it so much that the deep nuance of its original meaning has all but faded away, leaving only the superficial meaning. This metaphor has become almost a synonym for a simple plan. However, have you ever asked yourself why this metaphor was chosen in the first place? Very few people do so.

Think about the words that compose it. It is a map of the roads. A road is a linear structure. Granted, real roads tend to be winding, but you can reliably order the places of the points of interest on a single road in a sequence and measure the distance between them, not in terms of geometric proximity, but measuring along the road. A map is a different thing. A map is more than a single road. It is a network of interesting roads, where every junction represents a new point of interest.

Here lies the essence of the metaphor. A road is akin to a plan of execution. A list of places that you need to traverse in order to get from point A to point B is no different than a sequence of steps you need to take to reach a certain goal. You might avoid stopping in some places along the way, you might skip some steps, or add additional ones, but in general, there is no deviation from the set course.

Of course, if you encounter a roadblock or a traffic jam, you will get stuck. The achievement of your goal will get postponed, at least until the obstacle can be removed unless you can somehow pick another route and circumvent the roadblock. This is where you reach the map.

A map is a non-linear structure. Formally speaking, it is a directed graph. It contains the point at which you are, your starting point A. It also contains your goal, point B. It also contains all the possible roads and all the possible steps you might take in order to get there!

Having a roadmap provides you with the opportunity to choose an optimal path. Furthermore, it provides you with the opportunity to change your mind and pick another path that is more suitable to your current circumstances. It allows you to create a more flexible strategy, one in which you can course-correct if the circumstances change. This is the secret meaning of this metaphor that is so often lost in corporate slide decks!

Your spring is akin to a day on the road. Your resources are akin to fuel. Each day, you start with simple questions. How many miles can I cover today with the resources I have? How close will this get me to my goal?

Creating a roadmap doesn't mean listing one preferred ideal sequence of steps. Creating a roadmap should imply mapping out as many as possible roads and courses of action that could be taken. Some of them will be dead ends. Some paths will turn out to be blocked by insurmountable obstacles. It is also likely that you will encounter bottlenecks critical junctions that cannot be circumvented by taking another road. You might venture into uncharted territory, where you wouldn't be able to map possible paths in more than one or two steps into the future. However, in any case, you will always have more flexibility than a single road would offer, no matter how wide it is.

Live Service

9

Purpose of Live Service

Back in the day, things were simple. The team makes the game. It gets stamped on CDs or whichever other medium, gets put in a box, and shipped to the stores. The marketing team would plonk a huge amount of money on a promo campaign. Hype would rise until the launch day. People would buy the fixed price for the game and go home happy with their new purchase, to play the hell out of it for 5 min or 500 h. The team would have no clue about how much an average player would play their game, but they would not even care. They would just shrug, count the money, and go to make a sequel. Then someone had to come along and invent free-to-play, and now we all have to bother with things like retention and live service.

Recall Chap. 2 of this book and the discussion of various business models. Free-to-play and game-as-a-service business models have some objective advantages over more classical, i.e., premium model of the game business. Two of the most obvious ones are the following.

The game is distributed for free, for the price of 0, thus lowering the threshold for potential players to try it out to a theoretical minimum. In the classical capitalist framework, competition always puts downward pressure on prices. Competition in video games is extreme, so no wonder that the initial price of the game is set to zero.

On the flip side, from the player's perspective, this business model allows him to try the game out without making any initial investment save for the time needed to download it. This is reflected in the install bases of most successful F2P games, which routinely dwarf install numbers of even the most successful premium titles.

The other main advantage of this model is that it doesn't put a cap on the player's spending. A player might try the game for the cost of 0, but if happy, if entertained by the game, he will potentially continue to play and occasionally spend money on something

© The Author(s), under exclusive license to Springer Nature Switzerland AG 2024
S. Stanković, *Game Design for Free-to-Play Live Service*, Synthesis Lectures on Image, Video, and Multimedia Processing, https://doi.org/10.1007/978-3-031-56156-6_9

within the game. The player can repeat spending in the game whenever he feels like it. Thus, the total amount of money he would pay for the experience is not capped by the game's pricing model.

When launching and running a game as a service, there is a clear hierarchy of priorities.

The first priority should be the engagement of the players. If you get the players to install your game, the game needs to be entertaining for them to stick with it even for a while. Building engagement is within the domain of the core gameplay.

Once you get your players engaged with the game, you should strive to keep them engaged as long as possible in order to give them as much time as possible to decide to spend some money within your game. This is what retention is all about.

Finally, as profit is the ultimate goal of any business, monetization is your ultimate priority. One crucial thing to remember in the monetization of video games is that you must provide value for your players. Contrary to popular and naive opinion, there are no psychological ticks that can manipulate players into spending on your game. Your players are playing to be entertained, and they will not be doing so if they do not perceive value in what you are offering! Free-to-play players, in general, are very sophisticated customers. Indeed, they are some of the most sophisticated customers in the entire world of entertainment!

It should be noted that, in the text that follows, I do not make a distinction between cases when a player spends real money to purchase something in the game vs. the cases when the player spends hard currency to do the same thing. I treat both of these cases the same as they are ultimately closely coupled, both in the minds of players and from the design perspective. In addition, I will focus solely on monetization through in-app purchases and leave ad-based monetization discussions for another time.

Player retention and monetization are the domain of the live service. It is important to note that the same hierarchy of priorities exhibits itself in the launching of every new game feature or new live service event. For all practical intents and purposes, both new features and live events can be seen as mini-games bolted on top of a platform of your original gameplay experience.

In what follows, I will explore the interplay between retention and monetization that I have observed during the several years that I spent working on the successful live service. Do not treat this as a ready-made template. This is just a set of my personal observations. Rather, use it as a starting point in your thinking about the live service.

Live Service Strategy

I like to use a particular metaphor when discussing the live service strategy of any game. Ideally, a live service metagame should have the structure of a layered cake. Each distinct layer of the cake should be somehow built on top of the existing layers. All layers of the service should support each other and be geared toward the two primary objectives: player retention and monetization.

As your game progresses and evolves and, more importantly, as your audience and your knowledge about the motivations and behavior patterns of your audience grows, you should be able to add new layers to your metagame. Each new layer should have its own particular purpose and a reason for existence.

Contrary to popular opinion, in my experience, you do not actually need to have all the layers of your cake ready and implemented at the moment of global launching your game. There is a tendency of many game teams to try to have the complete feature set neatly implemented, tested out, and optimized during the initial soft launch period.

In reality, this can even be counterproductive. Above all, this requires both time and manpower to execute, making the team spread out its always scarce resources over a very wide front. Furthermore, before actually launching your game to the global audience, you are essentially operating on a set of assumptions. Things that on paper seem essential often end up being overtaken by other stuff once the game has launched.

Finally, business is all about growth. Your game needs to be able to stand on its legs, i.e., be profitable with a minimal set of features. If you have a concept that requires you to throw everything and a kitchen sink on it to make it marginally competitive, it is a sure sign of trouble up ahead. If you have already used all the tricks in the book to break even, how will you grow and scale up your business?

Instead, what you shout strive towards by the time of the soft launch are the following three things:

- A solid platform with engaging core gameplay that can support the ever-growing live service metagame.
- Minimal set of metagame features that support running a live service for an initial couple of months. My yardstick is around one to three months, depending on the type of game, but these numbers can be widely different.
- A clearly defined road map of metagame features that you would like to have in operation at the end of the first year of the live service. You should also be willing and ready to alter this roadmap and reprioritize the features within it according to the data that you will be gathering as you launch your game.

There are recent cases that can serve as an illustration of my points. Last year's indie hit, Among Us, is an example of a game whose success caught its makers by surprise. The game definitely launched with a minimal set of features to make it very successful, but it seems, at least from the outside, that the guys at Inner Sloth didn't have a plan prepared for running a live service, as shown by their dwindling numbers (Curry, 2023).

On the other hand, Zynga's FarmVille 3 launched with a full set of features after a prolonged soft launch time only to stagger in front of the global audience (Long, 2022).

Finally, the steady and relentless rise of Township by Playrix represents a testimony to the power of the gradual development of a live service (Katkoff, 2020).

It is also important to note that each layer of the cake should be able to stand its own legs independently from the rest of the edifice. In this way, you can afford to build one layer at a time, focusing your development efforts on optimizing the thing that you are building. This approach reduces the dependency between features during the development time while maximizing their dependencies once they are operational and reasonably optimized.

There is a tendency for various development teams to bite more than they can chew and build a very complex set of features before gaining the full understanding of the thing that they are making. Each separate layer of the cake needs to be validated in practice separately! If you attempt to ship a very complex and very interconnected metagame structure in one huge go, it is likely that you will not be able to tell which aspects of it are not working and why. Trying to optimize a dynamic system over multiple parameters all at once is not a sound strategy.

Focus on one layer at a time and get it into operational shape before moving to the next one, but always keep in mind the grand vision of the layered cake and the layer synergy!

Two Types of Monetization

When talking about the monetization of particular features, I like to make a distinction between *direct* and *indirect* monetization.

The first of these, direct monetization, is monetization in a narrow sense. Quite often, this is the type of monetization that PMs have in mind when they talk about the topic. This type of monetization occurs whenever a player decides to spend his money in exchange directly within your feature. The player might be paying to unlock a feature, skip waiting time, buy some resources or a gacha box, etc. Usually, in this kind of monetization, we expect the player to spend either real-life money or something directly proportional to it, for example, hard currency.

A premium pass is a typical example of a directly monetizing feature. A player spends a fixed amount of money to gain access to the premium reward track of the season pass.

Other features will not have such direct monetization opportunities, i.e., they do not offer anything that the player can directly purchase. Rather, they might motivate the player to spend his money elsewhere within your game. This is indirect monetization.

For example, you might have a limited-time event that offers as rewards attractive new content that won't be accessible once the event ends. To unlock these rewards, the player needs to play the core of the game and earn special event points. The game does not offer a direct way to purchase the missing event points, but the player has a chance to spend his money buying whichever items and resources, which might help him be more efficient at playing the core and grinding the event points to get the reward he is after.

Although indirect, such a monetization strategy can be at least as powerful, if not more powerful, than direct monetization. In addition, it has the advantage that it usually seems less greedy and more fair. In general, this strategy, if built well, can provide the opportunity to figure out the optimal ways to invest their money to achieve their goals within the game. Optimizing the spending pattern becomes a part of the game's learning curve! The very act of spending money thus becomes rewarding in itself. This is a very powerful emotion.

Consequently, both strategies should have their place in a well-managed live service.

Fundamentals of In-Game Economy

Before we get into the nitty–gritty of the monetization design, we need to take a look at some fundamental notions.

Actually, I hate the term monetization. It is quite unclear to me why it was even introduced. What we are doing here is nothing but good old sales. User acquisition is about marketing. Monetization is about selling! There is no way of overstating this. Above all, monetization is about creating value for your players. It is about presenting this value to the player.

It is surprising how often this is overlooked by novice teams. Monetization is not something you can slap onto your game. It is not something that can exist outside of your game. It needs to be an integral part of your gameplay experience.

There is no dark arcane craft to it. If your players see value in what you are offering, they will be happy to spend the money. Free-to-play players are a very sophisticated bunch of customers. They get to try your game and evaluate every purchase they make. If they are not satisfied with what they are getting, they will simply move away. There are plenty of other free games, just a finger tap away, waiting for their attention and their money. There is no way for you to cheat them. You can't make a successful business by bamboozling your customers.

In order to provide value to your players, you need to understand why players spend money in general. Recall the discussion from Chap. 2 of this book. It turns out that there are just three very general reasons why people spend their money:

- To achieve more, i.e., to be able to do something that they couldn't otherwise—for example, get into a concert venue to listen to their favorite band or drill a hole in a wall using the new tool that they just bought
- To establish identity, i.e., to express themselves—as a fan of their club by sporting a new jersey in the team colors or as a dapper gentleman in their new tweed suit
- To maintain relationships—by sending gifts on Christmas or flowers on St. Valentines

Any sort of items, such as power-ups, card packs, character upgrades, new weapons, etc., that have a gameplay value belong to the first category. Cosmetics, such as character skins, dance moves, emotes, etc., fall in the second category. The third category is about social spending features.

Furthermore, in order to establish value, what you actually need to do is establish the correspondence, i.e., exchange ratios between three very different sets of things. Number one is real money. This is obviously what your bottom line consists of.

The second variable represents the virtual goods you are selling. Your game should offer a variety of things to sell. However, in order to make your job easier, it helps to use one single virtual currency as a proxy for everything else. This is what hard currency stands for in games. It is sort of an interface between the real and virtual economy.

Finally, the third part of the equation is time! Time and money are in real life closely coupled resources. Money is actually a function of time. Quite often, players will trade money for time in your game. In other words, you need to establish how much money is one unit of virtual currency and one unit of the player's time worth in your game.

This being said, it doesn't mean that this needs to be the first step in building your game economy. Most often, it is a good approach to do some benchmarking and use the values from similar games as a starting point.

In addition, virtual currencies, like any currency, are prone to inflation. What is likely to happen is that over time, as your life service evolves, the value of some items and indeed, currencies will drastically change. You should be aware of these ratios whenever setting prices for anything in your game.

Sources and Sinks

Speaking of inflation, one last concept that you need to understand and map out is the concept of the sources and sinks of any currencies or types of items in your game. A source of items is any place in your game where any amount of a particular currency or any new item is added to the player's inventory. A sink is any spot in the game where an item or a bit of currency can be used, i.e., removed from the player's inventory. Ideally, the sum total of sinks and sources of each important currency would be in balance.

This especially applies to currencies and items that play an important part in your monetization. Double more so in the case of hard currency. Balancing sources and sinks is crucial. Making your sources of hard currency too scarce risks one or both of the following things happening.

In the worst case, players might perceive your game as being too greedy and unfair and start abandoning it, ruining your retention and any chance of monetization.

Players might start to value the hard currency too much and start hoarding the little that they have instead of spending it, leading to a stagnant game economy and ruining your efforts to actually monetize.

On the flip side, you might run a risk of being too generous with your hard currency. As expected, this can lead to people simply having too much of it, again ruining your monetization efforts.

There is a surprising tendency I noticed. It is also a mistake I made several times. Quite often, when introducing new features into live games, the game teams tend to try to entice players to try them out by offering very generous rewards. This can lead to the inflation of the hard currency. The feature should be fun and rewarding to play intrinsically without relying on extrinsic rewards for user engagement. The feature's FTUE (first-time user experience) needs to entice the players to engage with the feature. You should not have to bribe your players to interact with your features.

Another systemic mistake often made in the design of games is the pairing of infinite sources with the finite sinks of a currency. For example, if a player can earn a certain amount of currency, let's say coins, each day just by logging in, this source is infinite. In theory, a player can accumulate an infinite amount of coins just by coming and logging into the game. This is how taxes in SimCity BuildIt work. On the other hand, if a player is expected to use these coins to purchase a limited selection of other items, the sink is finite. This means that inevitably, once they have purchased all the items on offer, the players will start to accumulate the now-useless coins. The value of this currency, as with any useless commodity, will drop to zero. The same applies to any items that a player can produce at will and out of nothing in the game. This is why no farm game ever monetizes on stuff such as farm crops.

There is a lot of game design literature written on the topic of balancing sinks and sources of items in the game economy. See, for example (Emelyantseva, 2019). It is a huge and important topic. I do not intend to go into specific details in this particular text. I am just mentioning it to emphasize the importance of laying the proper foundations. However, I will mention one of the industry's dark secrets. Important as it is, the game economy balance is not the be-all and end-all of metagame design. Some of the most successful and longest-running games in the industry operate in an economy that is out of whack.

Strictly speaking, these foundations of the virtual economy are not part of the live service. However, they are the prerequisite for running an efficient live service. Our layered cake is built upon these foundations.

Evergreen Purchases

The bottom layer of our monetization cake consists of the oldest and the most venerable types of monetization features. These are the standard IAPs (in-app purchases) offered by almost all free-to-play games. Usually, these include packs of hard currency sold for real money and packs of soft currency that can be bought using hard currency.

The classical metaphors for these two types of currencies are Gold Coins as a representation of the soft currency and Gems (or diamonds) used to represent the more valuable hard currency. This is a common trope, but it is by no means universal. The advantage of reusing this trope is that it is familiar to the players. Since so many games use it, it has been established as a visual shorthand for the feature. By now, even the most casual player knows that Gems are more valuable than coins and that gems are usually purchased with real money.

Not all games employ two-tiered currency systems. Some employ only one (hard currency). The purpose of this dual currency system is to achieve flexibility. Hard currency represents a level of abstraction between real-world money and the game's economy. Games are normally operated in multiple markets. There are some 180 currencies used in the world today, each with its own exchange rate. The game teams need a level of abstraction between this myriad of real-world currencies and the rest of their game. The game economy can be designed and balanced using these abstract and universal gems as the fundamental variable. The pricing of the game packs can be adjusted separately without the need to tweak the rest of the game economy.

The consequence of having this additional layer of separation is that it becomes harder for players to establish the correspondence between the value of things on which they are spending gems in the game and the real money. This has been highlighted as one of the most ethically challenging aspects of free-to-play monetization. Keep in mind, though, that when designing such systems, one needs to, above all provide value for the customers. The old adage that you can fool some people all the time, and all the people some of the time, still applies. You can't build a successful business by trying to trick all the people all the time!

There are a couple of other things that one needs to keep in mind when designing monetization features at this level. By now, the industry has drifted into a recognizable template of having 3 or 6 packs of diamonds sold for cash and 3 to 6 packs of gold coins sold for coins (Fig. 9.1).

It is a good practice to follow, at least initially, this established template. However, one should not just blindly copy the example. It makes sense to understand some of the important details regarding these packs.

The pricing of the packs is by no means accidental. It has its own internal logic that should be understood. The first and the most important of all the price points is the price of the first, i.e., the smallest gem pack. This is the basic reference point that you will use to determine the prices of the rest of the packs in the store. This is (except in very special cases) also the cheapest thing to buy in your game. It also might be the first thing many of your players will purchase first. One might be tempted to put this price as low as possible to ensure a big conversion rate.

This price should never go below the *impulse purchase threshold* (Stefańska & Śmigielska, 2021). This is the amount of money people are ready to spend without batting an eye. This is the price of a small bag of Pringles or an ice cream in the grocery store,

Fig. 9.1 A screenshot of the store in Zooba by Wildlife Studios

this is the price of a bag of Pokemon cards or a cup of espresso. If you set it to anything below this threshold, you are leaving the money on the table!

Keep in mind, though, that impulse purchase thresholds can vary greatly from market to market. What is considered cheap in some places will be prohibitively expensive in other locations. Having different price points for different important markets might be necessary. This is where having a hard currency as a buffer layer comes in handy.

On the opposite end of the scale is the price of the biggest gem packs. One of the typical quips about free-to-play games is about the outrageous prices of items in the store. OMG, 100$ for a pack of gems! These huge gem packs, with their high prices, actually have a double purpose.

They are not necessarily intended to be bought by an average player. Above all, they are there as an *anchor point* to establish the range of values that the player might expect to see in this store (Sherif et al., 1958). In-game shops offer a range of virtual goods. They do not exist in the real world, and players might not have a frame of reference to evaluate the prices. Is 10$ a lot or little for these kinds of things? Is 20$? Is 50$? Setting a high anchor point helps establish this frame of reference. Some people will apply the strategy of buying the second most expensive or the second cheapest thing in the store. Selecting the appropriate price points for both ends is essential.

The players that do buy these high-value packs are obviously going to be big spenders or even super spenders in your game. These types of players have usually graduated to these high-value packs by purchasing something cheaper in the game first. This means that they will have a clear understanding of the value they are getting. They are also likely to make repeated purchases. They obviously have enough disposable income to spend on this game. If the price or the amount of goods that they are receiving is not large enough, you will get one of the most common complaints that this type of player makes. Why don't you let me spend money on your game? I want to spend this and this much money on your game, but you make me buy these small packs 10 times instead of buying only once!

We have already mentioned the correspondence between time, real money, and virtual currencies. This is one of the fundamentals of game monetization. In the same way, individual currencies within the game have their own exchange ratios. But here is the kicker: most often, you should not be using one single ratio, even for a single pair of currencies in your game. What you should be doing is encouraging bigger spending. One way of doing this is by offering a bulk discount. This means that when purchasing bigger gem packs, the player should play less per individual gem than when buying the smaller packs. The cheapest IAP is thus the least economical one. The same applies to packs of soft currency.

This inevitably complicates the balancing of your economy. The smaller packs tend to be bought by players that are early in the game, the bigger ones tend to be bought by players that have already invested a lot of emotional energy into the game. The value that they get out of them affects their own game progression. Be careful to take this into account when balancing all other game features that can be directly influenced by your currencies. Do at least two balancing models, one taking into account the best and one for the worst exchange ratio!

Convenience Purchases

As we mentioned before, both layers 1 and 2 include evergreen purchases. The reason why I separated them into two distinct layers is the context. Layer 1 consists of packs of hard and soft currency, i.e., gems and gold coins. These are something present in almost any free-to-play game. They are useful in any game regardless of the feature-specific context. For a player, they are a universal tool. A player can decide to purchase them at any point in the game with or without trying to solve any particular need.

The purchases that constitute layer 2, on the other hand, are always context-specific (Fig. 9.2). These include things like:

- buying of missing resources, either items or currencies, as found in many farming and city-building games,
- speeding up all sorts of timers,
- instant unlocking chests in games like Clash Royale, etc.,
- buying missing character shards, cards, or upgrade items in RPGs (Role Playing Games),
- buying additional energy or hearts in games employing energy mechanics, typically in Match 3 games, etc.

Fig. 9.2 Convenience purchases in Family Island by Melsoft Games

Various types of gacha boxes offered in the shop can also be seen as part of this monetization layer

These types of purchases are usually denominated in hard currency and can constitute major sinks of gems.

The trick for creating a successful indirect monetization feature is to make clever use of this particular monetization lawyer. A successful indirect monetization feature should motivate the player to repeatedly make various convenience purchases. The player should either face a puzzle that he is able to solve more easily by making several convenience purchases, or he should be in a contest with other players, which would offer him a chance to strategically use his money and other resources to gain an edge over the other players. In any case, in order for this to work properly, it must involve a significant level of decision-making on the player's part and a chance to make a mistake. Otherwise, the feature will inevitably have a pay-to-win feel to it.

Limited Time Events

Limited-time events are the bread and butter of live service. Coming up with new limited-time events is what your live team should be doing most of the time. I am talking here about the limited-time events in the narrow sense, i.e., events that have prescribed time duration and a significant game component.

The main purpose of limited-time events is not monetization. The main purpose of these events is player retention. Free-to-play games are all about routine making. The main, static part of the metagame is about building a regular daily routine for the player.

Fig. 9.3 The whole right side of the main screen of Family Island is crowded with limited-time event icons

Routines, however, eventually, by definition, become repetitive and boring, no matter how well crafted. Limited-time events are all about breaking the routine.

Ideally, a limited-time event would offer either a departure from the usual game rules or a new context for normal game activities. For example, Touchdown Tournaments in Clash Royale offer a new set of rules by which the game is played. These rules would likely not work as an individual game but offer a nice periodic diversion from the standard gameplay (Fig. 9.3).

Typically, a limited-time event will offer a player a reward or a range of rewards that can be won for a limited time only. By definition, the player has only a limited amount of time to qualify for the reward. Typically, a player needs to obtain a specific amount of points or currency in order to gain a reward. These points can be won by doing specific event-related activities.

Family Island events always feature their own unique event currency. These event points act as a distinct virtual currency. This has several advantages. Above all, the sources and sinks of this currency can be tightly controlled separately from the rest of the game economy. Second, this currency has only a limited time duration. It gets vaporized after the end of the vent. Both of these things ensure that this currency will not get inflated.

Offering more than one reward with different point requirements gives a chance to more players to win at least something and not leave the event empty-handed. It makes the balancing of the event easier, as players with various engagement levels can be rewarded adequately.

Rewards work best if they are a mix of unique new content, old content, and evergreen items and resources. In this way, various types of players have a better chance of finding a reward that is to their liking.

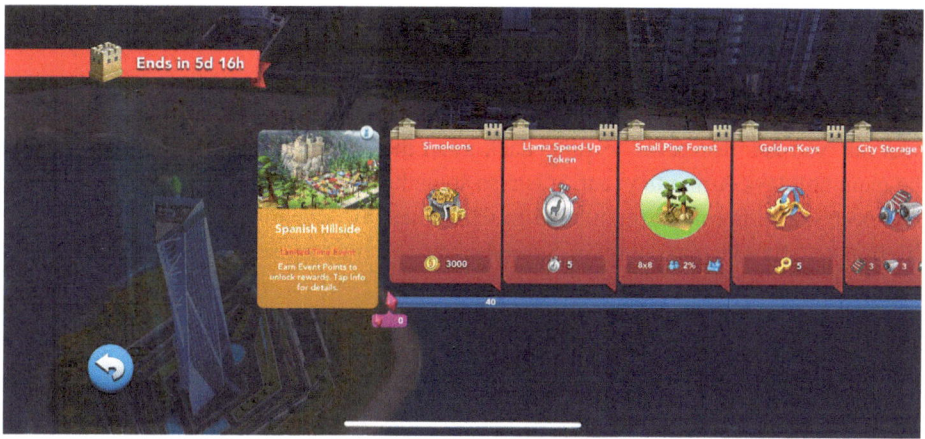

Fig. 9.4 An example of a limited-time event track in SimCity Buildit

Rewards can be presented in the form of a catalog with prices. This format has the advantage that the player can freely pick and choose the order of rewards, starting from the referred ones first. However, this can lead to players abandoning the event once they have unlocked the rewards they like, ignoring the rest.

Another way to present rewards is using a reward track, i.e., a list of reward tiers that a player needs to unlock in a sequence. The most enticing rewards can be placed on the end of the track, boosting feature retention. In this way, even an implicit narrative of scaling rewards can be built. However, this format can feel grindy and exhaustive to the players (Fig. 9.4).

From the monetization perspective, this type of feature tends to work best as indirect monetizing. The very notion that they are limited in time presents a challenge to the player. The player should be able to overcome this challenge by wisely using his resources and, consequently, his money.

To maximize the monetization potential, this type of event should be paired with the targeted offers described in the following sections.

Finally, one last parameter regarding the limited-time events is their duration and cadence. Ideally, in order to make your live game feel truly alive, there should be something new happening in your game every single day. limited-time events are only one of the middle layers of the live service. They are meant to work in conjunction with all other layers. You should aim at having at least one live event per month, if not one per week, in your live service. The cadence depends on the ability of your live team to produce the needed content.

The duration of these events is typically guided by the pacing of your core gameplay. If your game is relatively fast-paced and revolves around quick matches and short gameplay sessions, this should be reflected in the duration of the live events. Such games usually include live events lasting from one to five days. If your game is more leisurely paced, as is the case with the farm games, live events that last several days or even up to two weeks might be more suitable.

Tying your live events to the dates in the real-life calendar allows your game to stay relevant. You are also going to piggyback upon the general marketing hype built around the main holidays during the year. It is also something that the players expect. It is natural to expect something in the game tied to Christmas, St. Valentine's, or Halloween. When Starbucks starts serving Pumpkin Spice Latte, it is time to see carved pumpkins in the game.

Periodic Events

This live service layer consists of events that occur on a regular cadence. While the previous set of live events typically requires a significant amount of actual design work and hand-crafting for each new instance of the event, this category of events can be, to at least some degree, automated. In other words, the individual instances of these events follow an established template that can sometimes be triggered automatically.

Typical examples of such events are regular daily or weekly competitions. For example, both SimCity BuildIt and Hay Day run their own set of weekly competitions known as the Contest of Mayors and Derby, respectively. In both cases, these features are run fully automatically by the game's backend. Both of these features are indirectly monetizing, relying on evergreen purchases and periodic offers to deliver on the monetization front (Fig. 9.5).

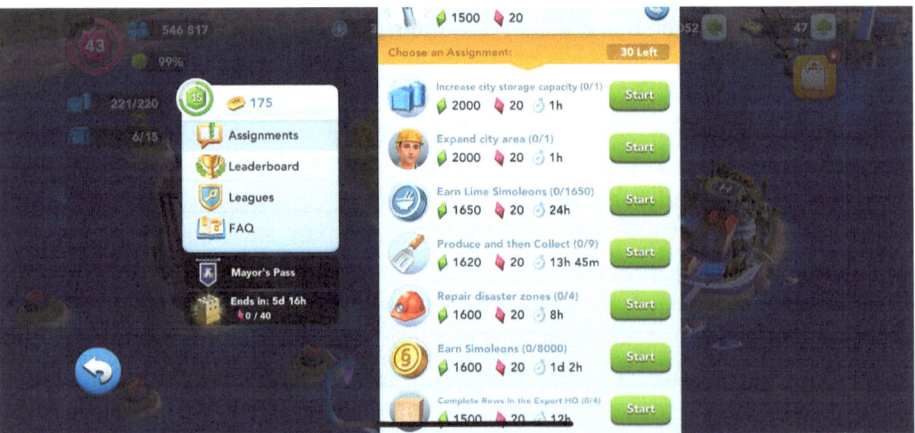

Fig. 9.5 Contest of Mayors in SimCity BuildIt

Fig. 9.6 Daily Goals in Hay Day, integrated into the Farm Pass UI

Daily Tasks or Daily Quests are another extremely common type of periodic event. This type of feature is one of the oldest in the free-to-play world. It had its origins even back in the premium days of games such as Jet Pack Joyride. In fact, daily tasks are so common that nowadays, even non-game applications like Duolingo have them. They are so ubiquitous that they are often overlooked. However, they can be extremely important. In many games, they serve to establish the regular daily context for interacting with the core gameplay. They are often the first step that binds the minute-to-minute interaction with the rest of the metagame (Fig. 9.6).

Since they require at least minimal engagement of players during the day, they tend to perform far better than simple login bonus features. While the login bonus rewards the player for simply starting the game, these tasks ask the player to actually play the game. Playing even one round of the core should remind the player of the joys of the game, leading back into the groove of gameplay.

Another typical and very popular example of such features is various incarnations of the premium pass mechanics. I have previously written about the premium pass features in detail. You might want to take a look at this text, Link.

Keep in mind that in most incarnations, premium pass features include a certain amount of new content. This means that they rarely run completely automatically and require some amount of handcrafting.

The true power of these periodic events comes from their periodicity. They provide regular time-boxed context for playing the core gameplay. A player has a renewed reason to revisit the core game each week over and over. These features provide the player with a set of goals to go after in each new round. As such, these features tend to stabilize both the retention and the monetization of the game.

Purchase Offers

I have reserved the offers for the last layer in this hierarchy, not because they are the least important. It is just the opposite. The offers are really where live service should deliver the punch. These are, by definition, the direct monetization features. However, their true power comes from the spending context created in conjunction with the rest of the metagame and especially with the rest of the live service layers.

There are several kinds of offers that free-to-play games employ, including:

- targeted offers,
- limited-time offers,
- periodic offers.

Each of these types of offers has its distinct purpose and should have its place in your monetization strategy.

Targeted Offers

As the name implies, targeted offers are focused on a particular subset of the player base. There are several ways in which offer targets can be selected.

Players can be targeted based on their progression within the game. For example, an offer can be triggered automatically when a player reaches a particular experience level or when a player unlocks a specific segment within a game, for example, a new arena. This type of targeting makes use of the context provided by the game progression. A player unlocking a new arena would benefit from buying a bundle of resources, allowing him to be competitive in the new level of competition.

Starter packs are one special type of such targeted offers. One key thing to remember about them is the moment at which they are offered. Starter packs are offered very early in the game and are specially designed to be the first purchase that a new player makes. At this point, the player most likely doesn't yet have a firm grasp of the game's economy. For most other target offers, the player is likely to already understand the context of the game. The player already has a mental model of the relative value of items and currencies in the game. Establishing a value proposition for a particular offer is, at least in theory, easier. The player will compare the value of every offer to other purchases in the game. The player will compare the value of the starter pack to his mental model formed outside of the game, guided by real-life experiences and other games.

Since the aim of starter packs is to give a boost to novice players and to normalize the spending in the game, they are one exception where going below the impulse purchase threshold is justifiable!

Fig. 9.7 An example of a targeted offer in Family Island

The players can also be targeted by other criteria. One typical example of targeting players is based on their spending profile. The evergreen purchases from layer 1 and layer 2 can serve as good guidance for this. A player who has already made several big purchases in the store might not be interested in a relatively cheap offer that yields relatively few resources. On the flip side, showing an expensive offer bundle to a player who has not spent at all or spent a relatively modest sum is not likely to produce results. Finally, offering a deep discount to players that are likely to make a full-price purchase of similar items is probably bad for business (Fig. 9.7).

Finally, players can be targeted based on some more idiosyncratic rules. Most games with a feature-rich metagame have a player base that consists of several distinct player segments. These segments can be characterized in terms of player interaction with one or a set of features. Once these clusters are identified through data analysis, distinct player profiles can be built for each of these clusters. Each of these player profiles would benefit from a different set of offers.

Limited-Time Offers

Limited-time offers represent a different category of offers. Again, as the name implies, these offers are usually tied to limited-time events. They are supposed to be the direct monetization counterpart of the indirect monetization limited-time events. The context in which they operate is created by the event itself. They should offer a chance to the player to wisely invest money in order to obtain the goals. For example, if an event is going to require the use of a particular character, offering a set of upgrade cards during the event might make sense.

Periodic Offers

Finally, periodic offers are a type of offers that are shown to players based on a fixed, predetermined schedule tied to the real-world calendar. The end of the month or the end of the week sales are typical examples of such offers. The advantage of these offers comes from their regularity. They tend to reinforce repeated purchase behavior. The most successful free-to-play games become digital hobbies for their most dedicated players. Such players tend to build a habit or a spending routine.—I love this game and spend 10€ on it each week. For that money, I get 50 gems and 20 power-ups. I use the powerups to win matches and gems to open loot boxes—. The sum of 10€ is really not all that much for a hobby; it's the price of 3 cups of coffee, but it comes up to more than 500€ per year. Repeated offers maximize this behavior of players.

On the other hand, the regularity is the main disadvantage of these offers. The players are smart. They will come to rely on them and avoid buying anything at full price if they can count on offers regularly occurring. If you are planning to integrate this type of offer, make sure to factor in their discount to the very basic structure of your game economy.

The Synergy of Live Service Elements

If you are working on a live game, most of the stuff I was talking about so far must seem painfully familiar to you. However, just stacking the layers of the live service one on top of another is not enough. The true magic starts happening if they start operation in synergy.

One of the most essential parts of organizing a successful live service is to find a way to make these layers work together and complement each other.

Examining the difference in the time horizons of various layers of your game is a good starting point. Your core gameplay is often called minute-to-minute interaction. These are activities that are expected to occupy the player's attention for several minutes, i.e., during one play session. On the opposite end of the spectrum, your metagame should ideally retain players for months if not years.

The life service part of the metagame is there to bridge this time gap! Here also, various layers play slightly different roles. Typically, limited-time events have a duration ranging from a single day to a week or two at maximum. This produces a nice set of medium-term goals for players to aim for while enjoying their minute-to-minute gameplay on a long journey through the metagame.

This pattern is further reinforced by periodic events. Ideally, these should have a regular rhythm and a time duration that again bridges the time difference between limited-time events and the long metagame. This is why these events tend to work best if they are timed to last anywhere from a week to several weeks. Premium pass seasons are generally designed to last between one to two months, with 6 weeks being often the norm!

Stacked on top of one another, these layers produce a timeline dense with goals that a player can pick and choose and aim for.

In the best case, you should make these events work together. If you are planning to run a sale offering premium currency or a bundle of items at a discount. It makes sense to schedule it so that it would overlap with an event that would open up a new sink for whatever you are selling. Creating an immediate need for the items that are on sale helps present the value of the purchase to the player. It also helps to offset the possible problems of the cannibalization of regular sales, as players are likely to spend a significant amount of stuff that they just purchased during the event.

Finally, in the ideal case, participation in one type of event should also provide players benefits in other events.

Consider, for example, a very common case of Christmas live events. Assume that the game has an ongoing weekly competition, thematized Premium Pass seasons, and a system of live events. To maximize the impact of the live service, the premium pass season that runs at the regular schedule and coincides with Christmas should bring in a set of new Christmas-themed content. It should also provide the player with some resources that can be used in live events. The live events happening in parallel should bring yet more content that thematically fits in with the stuff gained from the premium pass. They should also feature gameplay that acts as a sink for the items obtained there (Fig. 9.8).

This synergy can be further reinforced by having a special section in an item catalog or a similar feature, creating a mini collection that the player should complete by engaging with both types of limited-time events (Fig. 9.9).

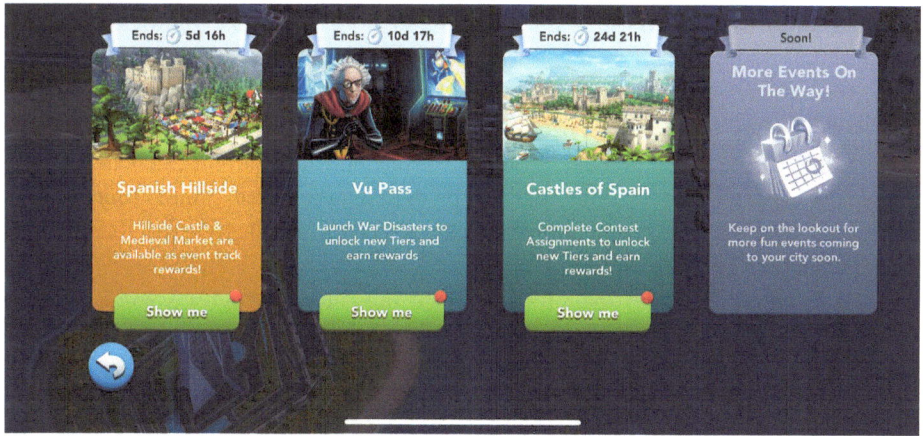

Fig. 9.8 An example of event Synergy in SimCity BuildIt—Spanish Hillside event and Castles of Spain premium pass season

Fig. 9.9 City Album page dedicated to Castles of Spain content in SimCity BuildIt

Furthermore, since both the Pass Season and the Christmas event are time-boxed, many of the players would benefit from having a sale offering suitable Christmas-themed bundles of items that can be used to gain rewards from the events faster.

This scheme creates a very dense and interconnected set of features. A player should ping pong from one layer onto another to gain a holistic gameplay experience. Various layers thus support each other, maximizing each other's benefits.

Live Service Calendar

In the previous series of sections, I have been talking extensively about the live service of free-to-play games. Limited-time events have a central position in any live game strategy. In this text, I will focus on practical examples of planning a live event calendar for a year of a typical game live service. Our goal will be to have at least one limited-time event per month, and ideally one every couple of weeks. Keep in mind that we are focusing only on individual live events.

These events will live to their true potential only if paired with a solid roaster of periodic live events such as premium passes or weekly or daily competitions and limited-time offers.

Retaining players to ensure monetization is the imperative of the live service, and that means delivering new content in regular time intervals. Indeed, this strategy is so prevalent in all contemporary games that players are grown to expect it.

Furthermore, In a well-run live service, limited-time events do more than simply deliver content in carefully measured portions. By their very nature, these events are time-boxed and ephemeral. This in turn creates a sense of urgency within the players' minds, motivating them to engage harder with the game in order to unlock the unique limited-time content.

Real-Life Holidays

In the virtual world of video games, there are no constraints about what the theme of a limited-time event might be. Anything imaginable is possible as long as it is within the capabilities of the development team. However, many games tie their events to real-world holidays.

There are some clear benefits to doing so. Above all, this fits well with the mental model that most people have. Players tend to get attracted to things that they expect to see. In addition, during the holiday season, exorbitant amounts of money are spent by companies in general on marketing that fits the seasonal spirit. In a world where every store window in the physical world is full of pink fluff and red hearts for the 14th of February, the collective marketing effort of the whole capitalist system blurs in one marketing campaign of gargantuan proportions, where everyone is piggybacking upon everyone else's advertising efforts. Therefore, it makes perfect sense to run this type of event inside your game.

Of course, global consumerist holidays are a no-brainer. Almost all games will have specific Christmas, St. Valentine's, Easter, and Halloween-themed events every year. However, if you truly want to maximize the potential of your live service, it makes a lot of sense to focus on the more culturally specific holidays as well.

If you are reading this text in English, chances are that your worldview is shaped along the lines of the so-called Western Anglo-Saxon mainstream. There is also a pretty big chance that this is actually not your native culture! Although we might be immersed in one unified global culture, that doesn't mean that the planet is monocultural. We live in a globalized, hyperconnected world. People are genuinely interested in other cultures. Your game probably addresses a very diverse audience. Seeing a holiday of your culture celebrated in a game creates an emotional bond. A good live service should cater to these needs.

Furthermore, first-tier partners like Apple and Google regularly organize the live service of their own app stores around the events in the real-world calendar. Over the years, the games I worked on enjoyed the benefits of getting regular store features around the holidays, which are less globally well-known. Getting an app store featuring, of course, lowers the cost of your user acquisition. That alone is the benefit enough to be worth your effort.

Calendar

In what follows, I will give an outline of a live events calendar featuring many holidays from different cultures from around the world. Some of these holidays have fixed dates, and others, like Easter, move around each year depending on sometimes complex calculations. This can be, at the same time, a challenge and an opportunity, as it allows you to shift certain events in order to better fulfill your calendar. I will present the events in a loose chronological order.

Lunar New Year

Date: anywhere between January 21st and February 20th.
  Region of Origin: China, East Asia.
  Description: Celebration of the start of a new year according to the lunar calendar.
  Iconography: The years in the lunar calendar are labeled according to the periodic 12-year cycle of animals in the East Asian zodiac, combined with a 5-year cycle of wood, fire, earth, metal, and water. These combinations repeat on a 60-year cycle. In practice, this means that each subsequent year will have a unique combination of iconic attributes that you can use to make stand-out content. The content your game use for the year of the Water Rabbit (2023) will definitely not be the same as the content you will roll out for the year of the Wooden Dragon (2024). Still, both batches of content will have the same unifying theme of the Lunar New Year.
  Colors: Dark red and gold.

St. Valentine's

Date: February 14th.
  Region of Origin: Global West.
  Description: Celebration of romantic love and affection. In Japan, on the other hand, this holiday is paired with another similar holiday known as the White Day, which takes place a month later. The modern tradition states that a woman should offer a gift of select chocolates to her love interest on February the 14th, while the chosen man should respond in kind a month later on. In Finland, it is known as Ystävänpäivä or Friend's Day, and the focus shifted from romance to a more general celebration of friendship.
  Iconography: Hearts, cupid with bow and arrow, ribbons, fluffy clouds, chocolates.
  Colors: Pink, white, red.

Carnival Season

Date: anywhere between January 28th to March 8th.
   Region of Origin: Brazil, Venice, Louisiana.
   Description: February is a carnival time in many countries. The tradition dates from the European medieval age. It marked the last party people would throw before the start of Great Lent. The best-known ones are:

- Rio Carnival (anywhere between February 1st to March 7th)
- Venice Carnival (anywhere between January 28th to March 4th)
- Mardi Gras in New Orleans (anywhere between February 4th to March 9th)
- Each one of them offers a distinct variation to the existing familiar theme.

Colors: Purple, Gold, Green, Turquoise.

Holi

Date: anywhere between February 17th and March 20th.
   Region of Origin: India.
   Description: Hindu Festival of Colors and Love. The festival involves crowds of people throwing handfuls of brightly colored fine powders at each other while singing and dancing. The festival begins on the full moon day (Purnima) in the Hindu month of Phalguna, which corresponds to February or March in the international calendar. The first day of Holi is known as Holika Dahan, and the second day is known as Rangwali Holi or Dhulandi, which is the day that the powder fights take place.
   Iconography: Bright colors! Puffs of brightly colored dust! Piles of colorful powders.
   Colors: Saturated shades of magenta, orange, turquoise, green, yellow, pink, purple, etc.

St. Patric's Day

See Fig. 9.10.
   Date: March 18th.
   Region of Origin: Ireland, US.
   Description: St. Patrick's Day is the patron saint of Ireland. This is a day of appreciation of the Irish culture and heritage. In addition to Ireland itself, it is especially popular in countries with significant Irish diaspora, including, of course, the US. In many cities on the East Coast, such as New York, Boston, and Chicago, this day is celebrated with a traditional parade.

Fig. 9.10 The splash screen of Family Island by Melsoft Games, advertising St. Patric's content

Iconography: Shamrock, Celtic spiral patterns, images of golden haps, ginger-haired leprechauns, and pots of gold, dark stout beer, see Fig. 9.10.

Colors: Green, also white, and orange.

Hanami

See Fig. 9.11.

Date: anywhere between March 20th to April 20th.

Region of Origin: Japan.

Description: Hanami is a celebration of spring focused on the admiration of Sakura, ephemeral cherry blossoms. The actual date varies from year to year and location by location following the blossoming of cherries. Japanese Meteorological Association actually creates a detailed forecast based on historical data and weather models. These forecasts are broadcast over local media. As warm air propagates over Japan, a wave of blossoming trees sweeps along the country from south to north.

Iconography: Cherry flowers and trees, traditional Japanese textile patterns. Figure 9.11 shows an example of Hanami content.

Colors: Baby pink.

Easter

Date: anywhere between March 22nd to April 25th.

Region of Origin: Europe.

Description: In cultures based on Christianity, it is usually the second most important holiday during the year, right after Christmas. The way various denominations celebrate it

Fig. 9.11 Hanami in SimCity BuildIt

can also differ significantly. Various Christian denominations calculate the date of easter according to various methods and formulas. In countries where the local church follows a different calendar, Easter may or may not fall on a different date. This includes Russia, Ukraine, Belarus, Serbia, Armenia, and even Romania.

Iconography: Painted eggs, decorated eggs, chocolate eggs, chickens, Easter Bunny, young green grass, and flowers. Local iconography can vary significantly from country to country, depending on the traditions.

Colors: Green, yellow.

Golden Week

Date: April 29th to May 5th.

Region of Origin: Japan.

Description: Golden Week is a series of holidays celebrated in Japan. These holidays are celebrated back to back, creating a week-long celebration. The first in the series is Showa Day on April 29, followed by Constitution Memorial Day on May 3, Greenery Day on May 4, and finally, Children's Day on May 5. This holiday is great for delivering a week-long event of Japanese-themed content.

Iconography: Japanese flags, Koinobori—fish-shaped flags.

Colors: Red, white, pink, yellow, gold.

May Day

Date: May 1st.

Region of Origin: Europe, US.

Description: International Labor Day manifestation. This holiday is deeply routed in the Socialist traditions. However, it is widely recognized and celebrated throughout continental Europe and in many places in the rest of the world. Having content suitable for this event can be a bit tricky as it is fundamentally rooted in leftwing politics. However, in certain countries, notably one influenced by Germanic culture, this holiday coincides with Walpurgis Night. In Finland, this is known as Vappu and is one of three major holidays during the year.

Iconography: In Finland: Helium balloons, doughnuts, mead.

Color: Red.

Ramadan

Date: May 23rd to July 23rd (Ramadan Eid).

Region of Origin: Middle East.

Description: Ramadan is a holy month in Islam. It is the 9th month in the Islamic lunar calendar. Observing Ramadan involves a strict fasting regime of avoiding any food or drink during the daytime and eating specific festive types of food during the night. Ramadan Eid is the celebration at the end of this fasting period and usually constitutes a proper feast. As the Islamic world is very diverse, local customs, food, etc., differ greatly from region to region and country to country (Fig. 9.12).

Iconography: A crescent moon, Arabic calligraphy, Arabic architecture, etc.

Color: Dark green, blue, indigo, silver.

Pride Month

See Fig. 9.13.

Date: June 1st to June 30th.

Region of Origin: US.

Description: This is an international celebration of the rights of LGBTQA+ people. It has its origins in the Stonewall riots, which took place on June 28, 1969. Showing support for the LGBTQA+ community is a norm in many countries of the global West, but note that it still might be illegal in certain jurisdictions. Even in the global west, in the US, for example, this can still be a hot-button issue that can lead to various reactions from different sections of the audience! Keep this in mind when planning your in-game events (Fig. 9.13).

Fig. 9.12 Example of Ramadan-themed battle pass

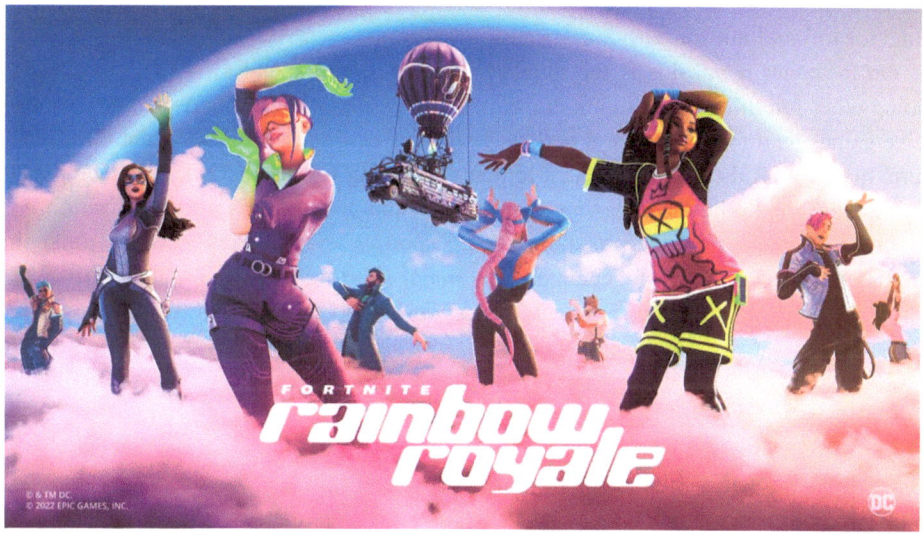

Fig. 9.13 Pride content in Fortnite

Iconography: Rainbows, rainbow flags, flowers, hearts, etc.
Color: Rainbow colors, pink, baby blue, etc.

Independence Day

Date: 4th of July.

Region of Origin: US.

Description: This is an American patriotic holiday celebrating the independence of the United States from the United Kingdom. This is a quintessentially American tradition, not observed in the rest of the world. Note it causes unintended reactions elsewhere in the world, especially in the places that have suffered from American neo-colonialism or interventionism.

Iconography: Fireworks, American flags, all things American.

Colors: Red, white, and blue.

Slow Summer Months

The time from the second half of July through the whole of August and September is, for some reason, almost totally devoid of big international holidays. There are many reasons for this. In ancient times, these were the months of intensive agricultural work. It was a time for toil and not for partying.

On the other hand, this is the time when most people spend a lot of time outdoors, doing other fun things, like sports, traveling, fishing, and doing all sorts of other hobbies. Live games tend to experience a period of slowdown where all the KPIs get hit. This, in turn, makes it all the more important to plan something for the live calendar. The priority at this time should be the engagement of players and not so much the monetization.

Luckily, summer is also the time of big sporting events. Piggybacking something sports-related onto the hype generated by a major sporting event can make sense. Watch out if the Olympic games or Football/Soccer World Cup is organized during the year.

Diwali

Date: anywhere between October 17th and November 15th.

Region of Origin: India.

Description: A Hindu festival of light. It celebrates the victory of good versus evil. It is usually a five-day celebration in the Hindu month of Kartika.

Iconography: Lanterns, lotus flowers, clay lamps, candles, flames, flowers, fireworks.

Colors: Red, white, and blue.

Halloween

Date: October 31st.

Region of Origin: Western Europe, US.

Description: The name of the holiday originates from All Hollow's Eve, a Christian celebration of All Saint's Day, but the actual origins of this holiday date back to paganic Europe and the Celtic harvest festival of Samhain. According to the legends, this is the one night in the year when the separation between reality and the netherworld is the thinnest, and the souls from the underworld can crawl out and roam the real world, thus the creepy connotations. Nowadays, the holiday is associated with pumpkin carving, costume parties, and trick-or-treating.

Iconography: Pumpkins, Jack-o-lanterns, candles, full moon, cobwebs, spiders, monsters, night, skeletons, candy.

Color: Orange.

Thanksgiving

See Fig. 9.14.

Date: November 22nd to November 28th.

Region of Origin: US.

Description: Thanksgiving is again one of the exclusive American holidays. It is essentially a harvest festival. The origins of this custom date back to the 17th century and the very first batch of English colonists that arrived in North America. Starting a colony on an alien continent is hard. In the first year, the colonists almost starved to death. They managed to survive only due to generous help given to them by the local native tribes. Thanksgiving is celebrated always on the last Thursday in November. This choice of timing helped anchor another important live service calendar event.

Iconography: Stuffed turkeys, pies, pumpkins, corn, and American flags. See an example of Fig. 9.14.

Color: Orange, brown.

Black Friday

Date: the Friday after Thanksgiving.

Region of Origin: US.

Description: Black Friday started off as a day when companies traditionally started offering discounts in a runup to the Christmas season. Nowadays, this is a celebration of naked consumerism. Things get sold on this day that would never be sold otherwise. Use

Fig. 9.14 Thanksgiving content in Family Farm

this day to run your most cash-grabbing events. All standards of decency get dropped during Black Friday.

Iconography: Sale signs, percent marks,

Color: Black.

Christmas

Date: December 24th (Christmas Eve) and December 25th.

Region of Origin: Europe, US (in its most commercialized form).

Description: This is, of course, the biggest and the most important celebration in the biggest part of the world. It is also arguably the most recognizable celebration worldwide. Although it has its roots in the Christian religion, it is now recognized and celebrated even in countries like Japan, which have very little to do with Christianity. Keep in mind, though, that not all Christian denominations celebrate this holiday at the same time. Some Christian churches still follow the old Julian calendar and celebrate Christmas on January 7th. In addition, in many parts of Eastern Europe, New Year's Eve can be a bigger and more important holiday than Christmas.

Iconography: Santa Claus, Christmas tree, Christmas ornaments, Christmas lights, gifts, mistletoe, etc.

Color: Red, white.

New Year's Eve

Date: December 31st, January 1st.

Region of Origin: Global.

Description: It is a celebration of the start of a new year according to the Gregorian (Global Calendar). Live events scheduled for this date are usually tied somehow to events scheduled for the Christmas season.

Iconography: Champagne, confetti, glitter.

Color: Blue, silver.

Practical Design Examples

10

First Time User Experience

Have you ever wished that you could skip a tutorial of a game you just installed? Have you ever dropped a new mobile game before even playing it once because it started to download 15 GB of something right after you tapped on an icon? Have you ever given up when a game demanded you open a new account on some service you have never heard of before letting you play?

Well, you have been a victim of a bad *First Time User Experience* (FTUE). If you are a game maker and want to spare your players from these kinds of torture, read on.

The Start

FTUE is a term used in the games industry to describe the initial contact of a player with a game. It is an acronym that stands for First Time User Experience, and it describes the process of user onboarding.

Many professionals in the industry tend to think of it in a relatively narrow sense. They equate the term with the idea of a *Tutorial*, a carefully crafted bit of design aimed at teaching the novice player the basics of the gameplay. Quite often, these tend to focus on the control scheme and functionality of the UI.

I prefer to see FTUE in a much wider sense. To me, it encompasses all of the interactions that a player will have with a game during the initial gameplay session. This can possibly extend to several subsequent sessions. The first-time user experience is something that the player will have regardless of whether you make an effort to deliberately design for it or not, so you better get prepared for it the best you can.

© The Author(s), under exclusive license to Springer Nature Switzerland AG 2024   227
S. Stanković, *Game Design for Free-to-Play Live Service*, Synthesis Lectures on Image, Video, and Multimedia Processing, https://doi.org/10.1007/978-3-031-56156-6_10

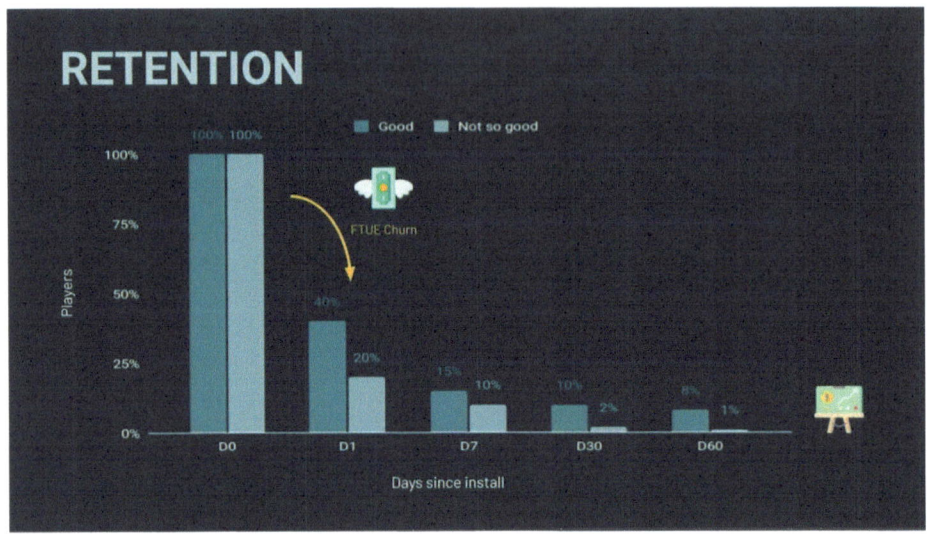

Fig. 10.1 Typical retention diagram of an F2P game

It is hard to overstate the importance of FTUE. If you take a look at the standard retention graph of any game, you will notice that the biggest drop in the number of active users happens exactly during the first play session. This is even more important in the case of free-to-play games, which are, as the name implies, available to players at zero cost. With premium games, especially AAA, the initial cost of the purchase that the player had to make serves as an additional incentive for the player to at least sit through the initial gameplay sequence, no matter how grueling it is. Not so in the world of mobile free-to-play, where play sessions are short, and attention spans are limited (see Fig. 10.1).

Designing a proper, high-performing FTUE is thus an art form in itself. Any mistakes made in this process can be very costly and difficult to fix. As with any art form, there is no foolproof universal method that can be magically applied to this task. However, in order to be able to create a good FTUE, one needs to understand what is its true purpose. The purpose of the FTUE is to create an emotional connection!

It is the 3rd decade of the twenty-first century. Video games have been around for more than 40 years, and mobile applications for more than 10. Most users are already familiar with the basic UI conventions. The users who gravitate toward your game are likely to be familiar with the types of interaction dictated by the genre of your game. Unless you are employing some novel and highly unusual control scheme, you have very little reason to focus your attention on handcrafting a tutorial that will explain these sorts of things. Don't make your player chase the yellow arrow!

What you should be doing instead is building engagement. You should try to motivate the player to make an emotional investment in your game!

The Journey

If you are familiar with the work of user acquisition specialists or growth hackers, you are most likely familiar with the notion of the Customer Journey. It is a way to describe the user experience that goes through several distinct stages, starting with awareness, the moment a potential customer encounters an app or an ad for an app for the first time, through consideration and installation, to retention, and finally, advocacy. In this framework, FTUE is situated between steps 3 and 4, installation and retention, with a distinct task of ensuring player retention (see Fig. 10.2).

This framework is very good for seeing the player's journey from the perspective of the business owner and product manager. However, to create a well-designed FTUE, one needs to go one step beyond and put himself into the position of the player to see the game from the player's own vantage point. To do so, it is useful to apply another journey framework.

The Hero's Journey is a concept proposed and popularised by Joseph Campbell in his seminal work on comparative mythology, a 1949 book titled The Hero with a Thousand Faces (Campbell, 1949). This concept, also known as Monomyth, has since become a darling of Hollywood scriptwriters and narrative designers everywhere. It postulates that most, if not all, of the epic stories in human history, share the basic underlying pattern, a sort of narrative blueprint that is shaped by our collective subconscious, shared genetic heritage, and evolutionary imperative. But what does it have to do with the FTUE?

Simply put, everyone wants to be a hero of his own story! Everyone wants to write their own success narrative. Your player wants to be a hero of your game. This is why he came here in the first place, so treat him as one!

Fig. 10.2 Place of FTUE in the player's journey

Call to Adventure

Every journey begins with the first step. Monomyth describes the journey of a hero as a series of distinct stages, starting from the familiar world and going into the unknown, the underworld, the land where the adventure takes place.

Just like any hero, your player begins his journey from the familiar setting of the home screen of his device. This first step is equivalent to the Awareness step of the Customer Journey.

The initial encounter can be initiated in several ways, for example:

- Organic conversion—when the player spots your game while browsing the App Store,
- Paid Ad—shown to the player in another game or on a social media channel,
- Viral—if the player learns about your game from an influencer, for example, via a YouTube or a Twitch channel,
- Word of mouth—when a friend or a relative recommends your game to a potential player.

Your task as a business owner and game creator, at this stage, is to convince the player to acquire, i.e., to install your game. An array of tools has been devised specifically for this task by the User Acquisition specialists.

In the context of the Monomyth, a hero can reject a call to adventure. Likewise, your potential player has the freedom to reject your game, even after he already installed it on his device.

To you, as a game developer, user acquisition is expensive. The cost per install has been growing on all platforms and will continue to grow in the future. On the other hand, the potential player has not invested anything to acquire a free game. This asymmetry is why retaining players is such a big imperative. You already paid the price of the player acquisition. You better keep them engaged, at least until you manage to convert them into paying customers! Keep in mind that not only do your ads need to be inviting to the player. Your game must also be!

Recall the SDT that we discussed in Chap. 3. Your game—especially its FTUE—must support players' psychological needs.

The first among these needs that comes into focus when designing an FTUE is the need for Autonomy, i.e., a need to be an active agent in one's own life. Video games are interactive by nature, implying a great degree of Autonomy in the playeržs actions. When a player faces a new game, his motivation is to get to satisfy his need for Autonomy as soon as possible. Everything standing in his way will be seen as a nuisance.

Avoid anything that would prolong the time before the player can get to do any meaningful interaction with the game.

A common error in FTUE design is violating this principle. Typical things to avoid include:

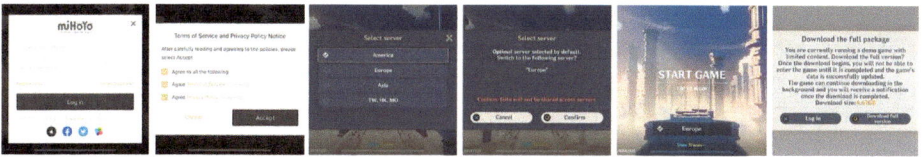

Fig. 10.3 Genshin impact manages to thrive despite an onslaught of pop-ups in FTUE

- Large initial downloads,
- Slow loading time,
- Mandatory logins using social networks,
- Requiring the player to create a new account on a new service,
- Intrusive pop-ups and dialogues.

Some of these things will be outside of your control. COPPA age gates, GDPR, Times of Service, Privacy Policy agreements, etc., are dictated by government regulations. However, although you are required to implement these intrusive UX elements, you do have a certain degree of control over the way they are implemented. Consolidate those as much as possible under the legal framework. Try to implement them, keeping in mind the overall guiding principle of trying to reduce the time before the player reaches the game's content.

To the players' subconscious mind, these intrusions read as if they were being asked to commit to something before even knowing what they are committing to. This is why they feel so bad! (see Fig. 10.3).

Other bits of FTUE are firmly under your creative control. One of the most common mistakes that game designers tend to make is to create an overly long section of dialogue. If you are trying to build a rich, intricate world, you might feel the urge to immerse your player in it as soon as possible, and you might imagine that the best way to do so is by throwing the information at your player from the get-go. You can be misled into thinking that you are building emotional engagement by introducing the elaborate backstory and a set of characters.

This is a mistake, akin to presenting a movie plot by drowning the viewer in the infodump exposition and voiceovers. Resist your narrative urge. Your player is not yet ready. He doesn't care if Arvena is in love with Lady Eyre, that there are no less than five moons orbiting around the world of Ae, and that the eternal Pillars of Urn have been handcrafted by wizard bards or wherever. Provide just enough of the backstory to pique his interest. Lead your player into the story by leaving a trail of breadcrumbs. Teach him about the game world by showing him. Provide just enough of the backstory to pique his interest. Lead your player into the story by leaving a trail of breadcrumbs (see Fig. 10.4).

The introduction to Taito's Bubble Bobble from 1986 is a masterpiece of minimalism. It consists of only three sentences. The first one promises excitement, helping establish the

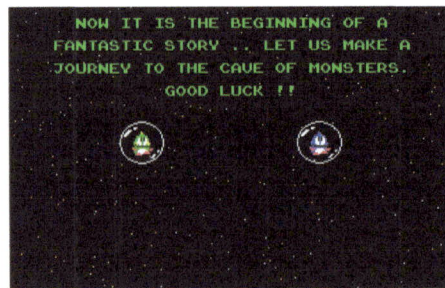 

Fig. 10.4 "Now is the beginning of a fantastic story!! Let's make a journey into a cave of monsters. Good luck!"—Bubble Bobble into

player's mood. The second one introduces the setting, mentioning the cave of monsters, taking almost verbatim from the Monomyth. The last sentence sounds more ominous. On one hand, it is full of foreboding. Why would one need luck unless one will face danger from those monsters, perhaps? On the other hand, it implies support. Someone is rooting for us. This intro provides zero factual information about the nature of the world in which the game is taking place. Yet it is choke-full with emotional hooks. What is best, you can grok it in a single glance!

Meeting the Mentor

The second step in the Hero's Journey involves meeting a magical figure, the one that acts as a guide into the unknown world. This stage of the journey is where the parallels with the standard FTUE practices are the most obvious.

In the Monomyth framework, this magical figure has two distinct roles. First, it is a figure of authority explaining to the hero the nature of his quest and the basic notions about the world in which the adventure will take place. Second, it acts as a gift giver, presenting the hero with one or more magical artifacts that will become useful later in the quest.

The use of a character in the game FTUE is one of the most established tropes in the games industry. This is usually either a humanoid figure or a creature that has been anthropomorphized enough so that basic emotions can be expressed and recognized by the player. Not all the games employ such characters in the FTUE. However, the advantage of using them over other simpler methods is backed up by the results of thorough psychological research (see Fig. 10.5).

The presence of humanoid features that are able to convey emotions helps establish an emotional connection between the player and the game. This emotional connection can be made by other means. However, the presence of the character with which we can empathize provides a useful shortcut.

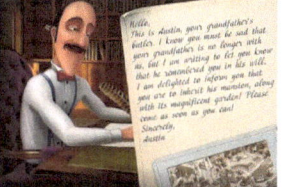

Fig. 10.5 Tutorial characters in Clash Royale and Clash of Clans by Supercell and Gardenscapes by Playrix

Many game FTUEs include a moment in which a player receives a set of items. These can be anything that makes sense within the given game: a weapon, a pack of cards, a health potion, a booster, a chest with randomized items, etc. By very definition, these objects, conjured out of nothing, are magic. This act of gift-giving is often even represented as such, i.e., as an act performed by the mentor character. If items are discovered by the protagonist instead, then the mentor at least provides the commentary, an interpretation of their nature.

Professor Willow from Pokémon GO by Niantic and Nintendo is a very good example of all of the aspects I described. The face of Professor Willow is the first thing that the player sees after the initial splash screen disappears.

This character both delivers the Call to Adventure and serves as a guide through the world of Pokémon GO. He is presented as an archetypal authority figure. His title, physical appearance, even posture, and relative height are carefully chosen to reinforce this impression. Professor Willow is there to reassure and entice. The player learns the basic notions about the world through a dialogue with him (see Fig. 10.6).

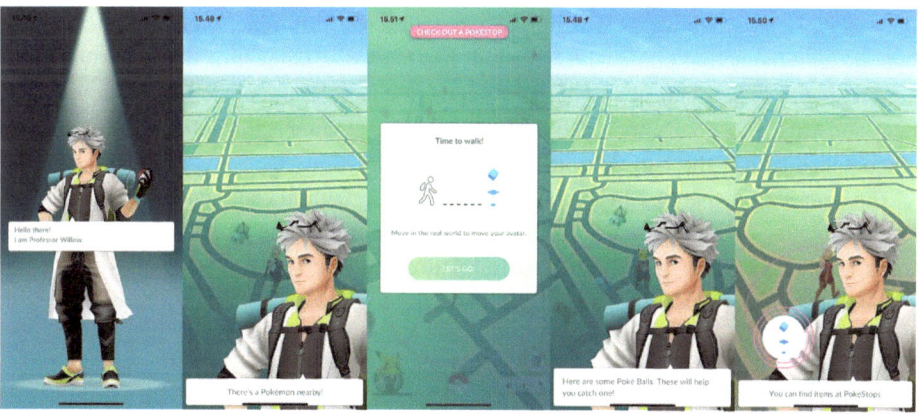

Fig. 10.6 Screenshots of some steps of Pokemon GO FTUE

Many of the items that appear at the core of the Pokémon GO gameplay are abstract creations without clear counterparts in real life. There is nothing resembling a Pokéball or a Pokéstop in the physical world. These are the quintessential magical items, even though they are depicted as technological artifacts. Indeed, even the Pokémon themselves are presented as a sort of mystical creature.

Professor Willow fulfills his role as a gift giver by donating the first set of items to the player, including a bunch of Pokéball, a backpack, an incubator, etc. He is also providing the interpretation of other magical items that the player discovers, such as the first encounter with a Pokégym.

Furthermore, this pattern is repeated throughout the game. The player continues to encounter Professor Willow in a series of mini FTUEs whenever a new feature or a gameplay mode is introduced into the game.

Crossing the Threshold

This is the third and final step of the Hero's Journey that we will devote our attention to. It is the moment in which the hero crosses the threshold between two worlds and steps into the unknown. It is also the act in which the hero meets the first obstacle and gets a chance to test his newfound skills.

In terms of the game FTUE, this is the moment at which the player finally gets to take continuous control over the game. This is a delicate moment!

We had already mentioned the basic psychological need for Autonomy. In the world of video games, equally important is the need for Competence.

The notion of Competence implies mastery of skill, i.e., the ability to correctly predict the outcome of one's own actions. Consider a simple act of kicking a ball. Our mind formulates our intention to kick the ball and premeditates the direction and the distance that the ball should move as a result of our kick. The difference between the intended outcome and the actual place where the ball has landed is a result of our skill. Obviously, if I am able to kick a ball in a way that it lends precisely at the place where the goalposts meet and trick the goalkeeper, I would perceive far greater satisfaction in my need for Competence than if the ball flew off in a random direction.

The first action that the player is allowed to take within your game needs to satisfy both needs, for Autonomy and Competence. The action that the player takes needs to feel meaningful in order to meet the need for Autonomy. Furthermore, the player needs to feel in control. The strictly regulated interactions presented as a part of a tutorial may not meet this criterion. What we are talking about here is the first action that a player can take by his own volition in a more or less unconstrained setting.

At this point, your player probably does not yet have a firm grasp of the context of your game. The action that he is required to take, therefore, must conform to his mental model.

Furthermore, as the player doesn't yet have full mastery of the game controls, the action needs to feel achievable by the player. An action that the player fails to perform will lead to frustration and can result in abandonment of the game!

To complicate matters further, it is very likely that the players of your game will possess very different abilities when it comes to actual gameplay. Depending on the type of game you are making and the audience that you are addressing, some of your players might be expert connoisseurs of the genre. They are likely to be well-versed in the design tropes and quite accustomed to similar control schemes. On the opposite end of the spectrum, you are also likely to encounter novice players—curious people who found your game through pure serendipity.

What makes things even more complicated at this point is that you, as a game developer, will have very little data about player behavior, even if your game is equipped with top-notch telemetry. Most probably, you will have no way to know what type of player you are dealing with in each individual case.

This wide spread of skills and prior knowledge of the game mechanics makes the task of designing an FTUE so much harder.

Recall the Theory of the Flow we mentioned in Chap. 3. You are essentially designing a one-size-fits-all user experience. For some of your players, this first task, or series of tasks, can feel trivial. They will find them boring. On the other hand, for some of your new players, the task might be frustratingly hard.

There is no surefire way to solve this conundrum. There is no game design silver bullet. Watching new users play through the FTUE of your game is one of the most cringe-inducing experiences of any game designer. It is about as pleasant as listening to the sound of the dentist's drill.

Rovio's Darkfire Heroes tries to teach the powers of no less than five heroes and two types of spells in the first three minutes of gameplay! There is no way for me to memorize and internalize all of these details. Instead, I rushed through the tutorial, tapping on spots shown by the disembodied hand, and skipped reading the text. For me personally, this experience was marked by boredom. The first level I played under my own control was marked by frustration. The tutorial I sleepwalked through left me utterly unprepared for the actual gameplay.

I don't know what their day 1 retention numbers are. However, most of this complexity could have been paced out over several smaller bits of strategically placed tutorials (see Fig. 10.7).

There are ways of mitigating these risks:

- Focus only on the bare minimum of actions that you really must teach players in order for them to enjoy the game.
- To avoid the chance of frustration, leave any advanced gameplay features for later. You are not trying to create expert players in the FTUE. You are supposed to teach them

Fig. 10.7 Rovio's Darkfire Heroes first-time user experience

just enough to have fun. There will be time to teach them these things. Or better yet, let them discover this stuff themselves.

- To further reduce the chance of frustration, relax the rules of your game as much as possible. If normally randomness would play a part in the gameplay, suspend it for the duration of FTUE.
- Provide immediate feedback on players' actions. Reduce any waiting times that might be part of your gameplay to trivial intervals. If they're bored by the actions, at least the players wouldn't be required to endure boredom for very long.
- Resist the urge to implement a skip button. An average person typically overestimates his abilities, especially if they are acquiring new skills! The player's decision to press this button is driven by his anticipation of the gameplay and not by his realistic estimate of his own skills. Allowing players to skip tutorials will, most likely, be a cause of frustration down the line once they hit the steep slope of the learning curve.

Don't let your hero die at the threshold!

Leaderboards

Since the arcade days, leaderboards have been one of the mainstays of video games. Even in a larger scope of the physical world, leaderboards are everywhere, from the UEFA Champions League Standings to the Little League Baseball (see Fig. 10.8).

If you are designing a game in which a player can accumulate some kind of score, chances are you will feel the urge to create a leaderboard. By definition, a leaderboard is an ordered list of players. It is usually presented in descending order, with the player with the highest score sitting at the top of the leaderboard.

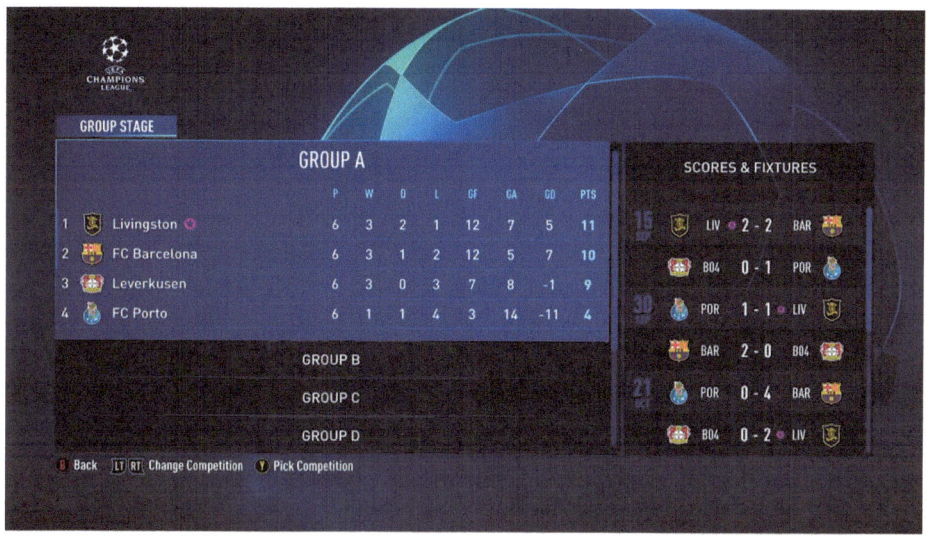

Fig. 10.8 An example of the UEFA Champions League Group standings

It might not be obvious at first glance, but leaderboards are a first stab at designing a metagame. They exist outside of the core gameplay. Even in the arcade games in the'80 s, they were seen on separate screens shown either before or after the actual game would start. Therefore we can safely conclude that the leaderboard is the start of a metagame!

A list is a simple structure, yet it holds powerful sway over the human psyche. For some reason, probably driven by the imperatives of our evolutionary survival, we are driven to compete against each other. A leaderboard does just that. It provides a means of comparison of our own performance at some activity against others. By some magic, it transforms a single-player game into a social competition!

Even this very rudimentary design offers a chance for players to instantly form several parallel goals:

- Reach the top of the leaderboard,
- If reaching the top is not immediately possible, at least try to climb up several places,
- If even this is hard, at least try to overtake the player immediately above you.

The time scale and effort needed to achieve these goals ranges in relative terms from immediate, overtaking the player immediately above, to long-term aspirational, reaching the top of the leaderboard. This is the nucleus of the Density of Goals I mentioned in Chap. 5.

This basic form of leaderboard does the job pretty well. As anyone playing any school game knows, these three goals can generate quite a lot of drama and excitement.

You might be tempted to take the design of leaderboards for granted and stay satisfied with this basic template. Yet this simple structure has a couple of fundamental weaknesses, which need to be understood and then taken into account if you want to maximize the impact of leaderboards in your game.

Pitfalls of Simple Design

The main problem with the basic leaderboard structure is that it doesn't scale well. There are two scaling issues that can manifest themselves depending on the type of game you are designing, time, the number of players, or both.

Time considerations originate in the speed by which players can accumulate points. If your leaderboard is refreshed after every core gameplay session (or even several times during a session), it can be the focus of players' attention. On the other hand, there is nothing worse than a stale leaderboard. If the time needed for any player to collect enough points to move up in the standings is too long, players will quickly lose interest in the leaderboard.

Think about the Eurosong voting. This is easily one of the most exciting contests on TV. Voting participants do not just give a single point/vote for the song they consider the most worthy. Instead, they get to distribute an increasingly larger number of points to several songs, starting from 1 and climbing all the way to 12. As a result, the standings on this leaderboard can change dramatically with each round of voting! (see Fig. 10.9).

Fig. 10.9 An example of the Eurovision Song Contest leaderboard

If you are designing a leaderboard, make sure to tie it to a score that can change significantly in an acceptable amount of time. A score that changes several times during one play session is a good starting point.

The second big issue in designing leaderboards is the question of scale, i.e., the number of players that can be simultaneously ranked in a single list.

Many games still contain some sort of a Global Leaderboard, i.e., a simple ranking of all the players that have accumulated any score within the game during a certain period of time. Given the number of players that even a mediocre game has nowadays, this list typically contains tens if not hundreds of thousands of names. This mistake is surprisingly common in game design, and it is hard to understand why.

As a player, I do not give a flying fuck about moving from place 156 342 to place 136 758 on some global leaderboard. The numbers are too large and too random to be understood on an emotional level. If anything, they are intimidating. The prospect of me reaching the Top 100, let alone the Top 10 on this leaderboard, is too daunting and too distant (see Fig. 10.10).

By virtue of things, there can be only one person at the top of the leaderboard. Good for him, but everyone else, including the guy at spot no. 2, will eventually feel frustrated.

Both the time and the scale problem are interconnected. They can work in tandem to exasperate each other. Some of these leaderboards are based on some kind of a slow-changing lifetime score. This gives a huge advantage to players who have simply started to play the game earlier, making the experience even worse for each new player joining the game. No matter how much time and effort I, as a new player, sink into the game, I will never be able to catch up with the advantage of millions of points that Top 10 players have accumulated simply because they had months or even years of a head start!

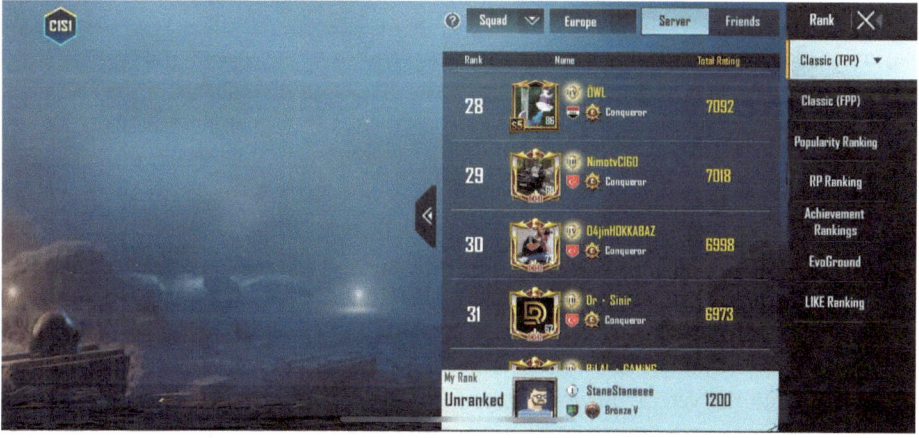

Fig. 10.10 Screenshot of PubG leaderboard for the region of Europe

The end result is a stagnant leaderboard with very little value for the players. The key question is why to have such a feature at all.

More Versatile Design

If we are aware of these issues, we can design around them and produce better, more meaningful leaderboards.

The first step is something that I already mentioned. The leaderboard should be tied to some relatively fast-changing score, ideally one that can change during a single gameplay session. This will go a long way toward creating emotional suspense. To achieve this, find a way to tie your leaderboard to a fast-changing score.

The second step has to do with the scale of the leaderboard. Moving only one or two places on one leaderboard can feel infinitely more meaningful than moving 50,000 places in another. Divide your players into several small groups. The group size can vary and depends on the type of game. However, it shouldn't be smaller than 10 or larger than 200 players. In my experience, a group size of 100 is a good number. Each individual group should have its own separate leaderboard instead of having a global one. These tight groups, in combination with fast-changing scores, should provide for intense competition.

Numbers from 1 to 100 are small enough to feel intimate. Their meaning can be grokked immediately. Moving a couple of places on this small leaderboard feels intuitive and meaningful.

This leaderboard structure works well with league ladders, a feature that we discuss in the next section of this book.

Finally, you can overlay a reward scheme on top of this system. Reward players at the end of each round based on their standing. For example, the top player should get the Grand Prize, players in places 2–5 should get a Big Prize (whatever that might be), players in places 6–20 can get a Mediocre Prize, and everyone between 21 and 50 could get the Consolation Prize. Players ranking below 50 get diddly-squat.

The Density of Goals in this structure is dramatically bigger than in the case of a simple leaderboard. The player can pick and choose his next goal based on his abilities and current standing on the leaderboard. He can decide to try to overtake another player, move up the leaderboard, or reach the top position. In addition to these, he can choose to reach the promotion threshold for the next league. He can also aim to move up the league ladder or eventually reach the top league. He can even dream about reaching the top place in the world.

Since scores reset at the end of each competition round, everyone has a fair chance of catching up with the veteran players. Players can also aim for a particular reward bracket within their current leaderboard. At least they can try to avoid being demoted to the league below.

Keep in mind that in this context, climbing the leaderboard and a league ladder is a form of intrinsic motivation, while rewards granted are a form of extrinsic motivation. You can read more about different types of player motivation in my previous text.

Limitations

Of course, the structure that I just described is much more complex than the simple leaderboard. This complexity needs to be taken into account when planning the development of such a feature.

However, in addition to the higher development cost, this approach has some conceptual issues. What happens if the player reaches the top position? Well, OK, good for them. He is going to feel awesome for a little while. He might even feel the pleasure of fighting to maintain his top position. Also, each new victory will only yield diminishing returns. The game essentially ends for this player.

Furthermore, the difficulty of the competition within each group depends on the skills and abilities of all players competing in each group. The criteria used to form the competition groups, i.e., for matchmaking, is important. Ideally, competition should be fair. This means that players with similar skill and engagement levels should be grouped together. This guarantees the toughness of the competition but not necessarily its fairness in broader sense. Players competing in tougher competition groups will need significantly more effort to reach the top of the leaderboard than players in easier groups. The rewards need to be proportional to the effort. You can, of course, always give bigger rewards to some groups of players and smaller to others. However, without some sort of underlying structure, this might seem arbitrary to players.

As mentioned before, one way to solve some of these problems is to build a system of leagues on top of your leaderboard competition structure. In practical terms, this means creating another metagame layer.

League Ladder

As we mentioned in the previous section, league ladders represent a natural extension of the leaderboard systems.

You can construct a ladder of leagues directly on top of the leaderboard system. Periodically, for example, after each competition round, you can promote the top-ranking players from each of the groups into the next league. You can also downgrade the players from the bottom of the group to a league below.

Since, in each iteration, only a fraction of players from each competition group get promoted to the next higher league, the number of competition groups gets smaller the higher we go up the league ladder. This creates a pyramid structure. Based on your player base numbers, you can balance the number of leagues and the size of groups so that the top highest league has only one competition group. The winner of this group will, in turn, be a world champion of your competition.

Leagues, in effect, act as a filtering mechanism, naturally separating the players based on their skill or engagement levels. Their true impact is often felt only in the top few leagues. The competition in bottom leagues tends to be very easy and uneven the higher the players get the opponents will be better matched to their skill level.

Ideally, the system should be implemented in such a way that you can tweak the number or percentage of players that get promoted or demoted for each league separately. This is important for several reasons. First of all, once you deploy this system by definition, all players will be competing within the bottom league. Competition at that level will be relatively boring for a lot of players. You might want to promote a bigger percentage of players in order to move them faster to a higher, more competitive league in order to populate these leagues faster.

The same observation is true for each new player joining the competition for the first time. As the actual skill and engagement level of a new player is unknown, the player will naturally start at the bottom of the ladder. Typically, the expected place for the new player will be somewhere in the middle of the ladder. By promoting a bigger percentage of players in the bottom leagues, you ensure that the players will reach their expected rank faster.

Furthermore, climbing the ladder is fun. Dropping down is not. You may want to reward players for doing well by demoting fewer players than promoting. This will make losing less punishable and winning more rewarding. Climbing the league ladder is, in this case, easier and, arguably feels better to the players. As a consequence, your pyramid structure might become disbalances. It is expected that the skill level of players will follow a normal distribution. Unless you have a steady influx of new players to populate the bottom leagues, most of your players will eventually make their way to the middle leagues. Your promotion and demotion shenanigans may lead to a deformed pyramid structure. The influx of new players and the distribution of players in leagues is something that needs to be constantly monitored!

Finally, the accumulation of players in the middle leagues can lead to stagnant competition and boring gameplay for a majority of players. A player who is stuck in league 3 in the system with 5 leagues will be likely to eventually lose the motivation to play.

On the other hand, what happens to the players that reach the top league? Reaching that level of competition is rewarding. Reaching the very top might be an ultimate challenge, but what happens if a small number of elite players attain the top places in the top leaderboard? Their very presence is a blocker to other aspiring candidates to do so. There needs to be a mechanism that ensures that competition, even at the top level, remains interesting. The number of players in this elite group might be tiny compared to your overall player base, but these are most likely your most ardent players and best customers. Your game needs to cater to their needs.

One small thing you can do is to make falling out of the top leag easier than reaching it. You can always demote a relatively large number of players from the top leagues down to the earlier ones. This ensures a constant flux of fresh blood in this part of the system.

However, it doesn't solve all the problems. By design, your system will have a finite number of leagues, be it 5, 9, or 24. The players, especially the highest engaging ones, would like to constantly be climbing the league ladder ad infinitum.

One way of solving this is to periodically reset the league ladder and the scores of all players. This would result in all players having to reclimb the league ladder all over again, reliving the complete user experience. This has a disadvantage in that players would lose their top league status each time the ladder gets reset. This is a hard sell. To make it easier, ensure that this is part of the game from the very moment of introduction of the league ladder. Introducing this system later is definitely possible. However, it is much harder to explain to players than if it would be there in the first place.

The additional twist that some games employ is that the reset is never complete. Instead, only the top, or top few, leagues are reconstructed periodically. The players from those leagues are relegated but not to the very bottom of the ladder. Rather, they are placed in some of the middle leagues according to some criterion. For example, based on their leaderboard standing at the end of the reset period.

Finally, when constructing a league system, keep in mind that eventually, you might need to extend it by adding more leagues on the top of the ladder. This might lead to both conceptual and technical problems. Technical problems are related to the backend architecture. The conceptual problems might be related to, among other things, the naming of the leagues. If your top league is already called 1st League, how do you name a new league that goes above it? If your top league is already Gold League, is the new topmost league going to be called Platinum? What if you need to add yet another one? Diamond? Palladium? Adamantiom league? Unobtainium League? Super duper ultimate AAA+++ league? Keep this in mind when constructing your system.

Premium Pass

The business of mobile free-to-play games is a ruthless, merciless one. Hundreds of thousands of new games each year fight for the attention of millions of fickle players, spoiled by endless choices offered to them.

Winners can make millions, even billions in revenue, yet to make a big fortune in this market, you need to start with a small one. Powered by the war chests of accumulated loot, established heavies lock horns with ambitious upstarts, spending millions on advertisement of their games.

With all this money at stake, it is no wonder that the game design of these games is evolving at a crazy pace, especially when it comes to metagame design, the bread and butter of the free-to-play business.

Every couple of years, a new metagame format pops up and swoops the hearts and minds of game designers. From Saga Maps and 4X to Card Collections, new concepts arrive from nowhere, and then suddenly it looks like everyone is copying the pattern, afraid to miss out on the next big thing.

The Premium Pass mechanic is one such case. Now, in the summer of 2021, it seems that almost every successful free-to-play game has its own implementation of this structure. If you are interested in what makes this mechanic tick, read on.

As a game feature, this concept came into being with the sudden advancement of Battle Royale games. These are inherently multiplayer games. The basic premise is that each participant in the combat starts each match from zero, thrown weaponless into the wasteland. None of the participants thus enjoys an unfair advantage over the others. During the course of a match, participants are expected to forage for weapons and pieces of protective gear, augmenting their capabilities.

This setup makes the design of a meaningful metagame very difficult. The challenge that the designers of these games were facing was how to build a metagame progression that would feel compelling to players but would not destroy the aforementioned basic tenets of the core gameplay. Offering a simple weapon upgrading system would surely violate the very idea of the genre, and without a well-developed metagame, how would you expect to retain, let alone monetize your players?

The solution that they managed to devise is what we now know as the Premium Pass mechanic. Therefore, the first thing that we need to acknowledge is that in the Battle Royale games, the Battle Pass system was designed first and foremost as a retention mechanic! To be sure, the very existence of this metagame backbone allows for the monetization of these games.

As the power of this system gradually got recognized, it got adapted to a multitude of other game genres. In some of those, which have other stronger metagame structures, the main purpose of the Battle Pass evolved more into the direction of direct monetization.

The versatile and flexible structure of the Premium Pass systems allows for this dual role. The capability to act in these two capacities is what makes Premium Pass systems so attractive to game designers and product managers.

Structure

The key part of every Battle Pass system is the reward track. It actually consists of two parallel tracks, one available to all eligible players, usually to everyone playing the game, and the other one accessible only to the players that had made a purchase of an access ticket. This ticket is usually known as the Premium Pass.

Reward tracks are subdivided into a great number of reward tiers. Players gain access to individual tiers by completing some unlocking criteria. The most typical unlock criterion is accumulating a certain number of points or some sort.

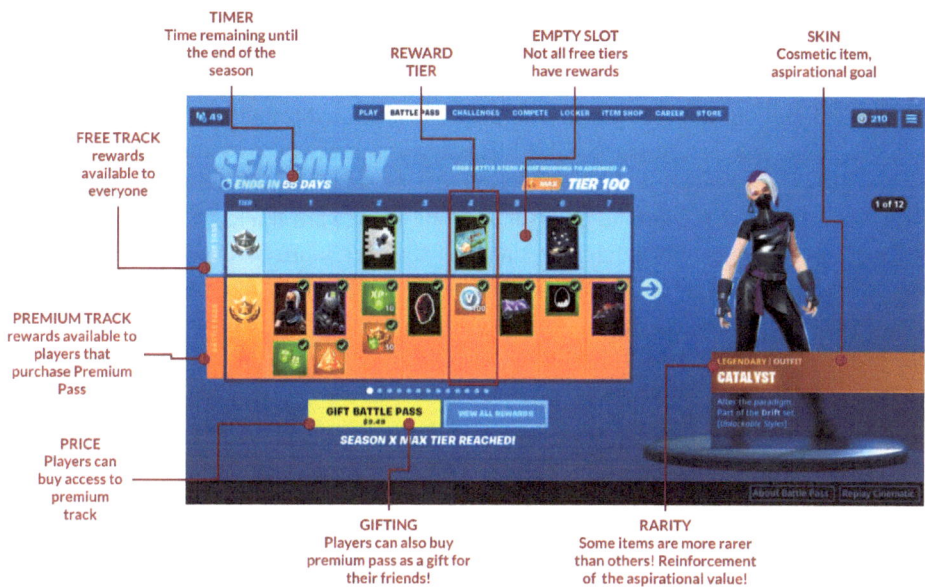

Fig. 10.11 Anatomy of a battle pass system

A player can claim a reward from a particular tier after unlocking the tier. In other words, a reward from a particular tier in the Free reward track can be collected by any player that reaches a certain point threshold. On the other hand, in order to collect a reward from a particular tier in a Premium reward track, a player needs to both accumulate enough points and purchase the Premium Pass.

Access to the reward track is limited for a certain period of time, usually called a season. The content of the reward track gets renewed with each season, but so is the player progression. At the start of each season, the scores of all players get reset to zero. Players are required to buy a new Premium Pass to access premium rewards (see Fig. 10.11).

Mechanics

There are some subtle aspects of the design that make the Battle Pass systems so effective and popular among both the players and the product managers of the game industry.

The way in which players collect the points needed to unlock the reward tiers is usually by playing the core of the game.

A key design principle that underlines this type of feature is that points are gained by participating, not necessarily by winning. The players can gain points for making kills in a Battle Royale match or for causing damage in MOBA-style combat, or simply by doing any other core gameplay activity. The player is not required to actually win a match or clear a level. Winning can still bring in a substantial bonus, but it is not a prerequisite.

In this way, the Battle Pass mechanic separates the player's progress in the reward track from his progress in the core game. The player is rewarded primarily for his engagement with the game and less so for his skill! This is what makes the system so attractive for the players. A player who is stuck on a particular level or keeps consistently losing battle royal matches can still continue to advance along the reward track.

This largely annuls the "unfair" advantage of stronger, more advanced players in Player versus Player (PvP) games. It doesn't matter that I was a Level 11 player and that I am losing to Level 13 players as long as I am able to unlock new rewards from the track. They can have fun trashing me, the same way I am trashing the Level 9 players. We all walk away with some points. A 1:1 PvP game is not a zero-sum game anymore!

In PvP games, the player is rewarded for his commitment to the game and his engagement. His winnings are not influenced anymore by the skill of his opponents!

The time box that the season structure imposes adds to the sense of urgency. A player tries to unlock as many reward tiers as he can in the allocated amount of time. This creates a positive feedback loop with player engagement. The more I play, the more rewards I will unlock. The more rewards I unlock, the more I will be motivated to play.

Furthermore, a player can collect rewards from the free track without making any purchases. He is at each step of the way tempted by the rewards that he might be collecting if he purchases the Premium Pass. In most incarnations, the system allows the player to make a purchase at any moment during the season. The players that buy the Premium Pass later during the season are still allowed to retroactively collect the rewards from the premium tiers that they had already unlocked based on the number of points they have collected!

This creates another powerful incentive. A player can decide to convert to the Premium tier only after he has unlocked a substantial number of reward tiers. In other words, he can spend money only after he is certain that he will get his money's worth of prizes!

Usually, Premium Pass tiers are balanced so that the players will get a fair amount of prizes already after unlocking the first part of the Premium reward track.

The further down the reward track player progress, the greater the incentive for him to convert. Since players are required to buy a new pass for each season, the players that consistently participate in them will know what to expect, how far they will be able to reach, and the amount of effort they are willing to commit to. The value proposition is self-balancing! Players can buy the Premium Pass only after they are sure they will get enough value for their money!

For the players who opt to buy the Premium Pass at the start of a season or in the first few days, the purchase itself serves as an incentive to play. They will try to get as many rewards as possible in order to maximize their return on the investment. Of course, the more they engage with the game, the more value they will derive from their investment, ultimately making their own user experience better. If bought early on, the Premium Pass acts as an investment, motivating the player to engage further.

Limits of the System

The Battle Pass system might look like a holy grail of free-to-play metagame design. It is a system generally loved by both the players and the product managers. It rewards the players for their efforts rather than for success or luck. However, it, too, has its limits. This is something that some of the game design teams have discovered the hard way.

The key to the success of any Battle Pass system is the value proposition. The perceived value of the Premium Pass to the players depends on the value of the rewards in the Reward Track.

In general, there are two types of rewards that you can put in a reward track:

- Cosmetic or vanity items—such as costumes (also known as skins), hats, pets, emoticons, victory dance moves, etc.
- Items with direct gameplay value—new weapons, various power-ups, cards or packs of cars in Card Collecting Games, upgrade tokens, etc.

Cosmetic items serve the purpose of establishing and reinforcing the player's identity and allow him to boast about his previous success. Look, I have this silly banana suit because I was able to unlock tier 35 in Season 3! I am that good! In order for these items to have value, they require an audience. They are, therefore, best suited for multiplayer games where the player's opponents constitute this audience.

Even if your game is based on this type of gameplay, you are still facing a logistic problem. The rewards are usually permanent. In order to maintain the success of your Battle Pass system, you are forced to provide players with new content each season. This is the dreaded content treadmill that so many free-to-play teams try to avoid. After having the plain old banana suit, you will end up with a green banana suit in Season 3, a pink banana suit in Season 4, a golden banana suit in Season 7, and a super deluxe rainbow glitter banana suit in Season 14 (see Fig. 10.12).

The items that have gameplay value have their own set of risks. Ideally, these would be consumable items that players would constantly find useful in the core game. In order to have these, you need to have a robust metagame that already supports such items and that can be deep enough to sink them. This is not the case in all genres of games. If your game supports either of these, a Battle Pass system will not have a great chance of success.

Fig. 10.12 Golden version of Agent Peely banana skin from Fortnite

Piggy Bank

The first of these features is known as the Piggy Bank. It is actually quite often, but not always, even presented using this visual metaphor. This feature works in the following way.

A certain amount of virtual currency, either soft or hard, is added to the piggy bank whenever a player achieves something within the game. This goal can be anything, from reaching a particular level to opening a lootbox or even winning a match. This is presented as the player as if he is earning the currency.

The player doesn't have direct access to this currency. However, the player can choose to open the piggy bank at any point. Opening the bank requires making an in-app purchase.

Usually, the bank has an upper limit, i.e., the maximal amount, after which no new currency can be added. Typically, the IAP is priced so that the currency is actually sold at a discount. This value proposition is usually communicated to the player as part of the piggy bank UI. The psychological angle of this feature is that the player has earned the discount through the invested effort.

The piggy banks can sometimes be limited-time purchases or even periodic ones to capitalize on the fear of missing out impulses.

Clash Royale offers a typical example of this feature. The Piggy Bank is a limited-time feature that costs about 6€ to open. All the gold that the player collects while the feature is active gets added to the piggybank with a $3 \times$ multiplier. The soft currency is thus sold

Fig. 10.13 Three little piggies in SimCity BuildIt

to the player at a heavy discount. The bank has both upper of the amount of gold that can be stored and a minimal amount that needs to be in the bank for the bank to be open (see Fig. 10.13).

SimCity BuildIt has a very sophisticated version of this feature, tied to a bigger feature known as Regions. What sets it apart is that it includes actually three offers that a dedicated player can unlock in the sequence. It thus offers three price points with different value propositions to the players, allowing for more flexibility.

Tower of Fortune

The last feature that I am going to examine is known as the has been developed by Angry Birds 2 by Rovio. I have insider information that the introduction of this feature has made a significant impact on the KPIs for this game.

The feature is built around the metaphor of an elevator. The player starts from the ground level and tries to climb to the top of the tower. The elevator stops on every floor. Whenever an elevator stops, the player is presented with a set of cards.

The content of the cards is naturally hidden, and the player is required to choose one of the cards. Most of the cards contain some goodies, such as items or amounts of various virtual currencies, hard or soft. One of the cards hides a bomb (presented as a Bad Piggy).

If the player picks a card with a reward, the reward is added to the player's winning pot. The player can choose to get off the elevator on any floor, i.e., and collect the winnings. If the player draws a bomb, the game ends, and all the winnings are vaporized.

Fig. 10.14 Tower of Fortune in Angry Birds 2

Each stop of the elevator thus becomes a wager, and the player needs to decide if he should take the risk of losing all that had been won already in order to increase the total winnings. It is a classical dilemma that skirts the concept of gambling.

As expected, the rewards become more lucrative the higher the player gets. The player's chances of dodging a bomb drop with the number of floors that the player climbs. Both risk and reward become higher the longer the game continues (see Fig. 10.14).

Of course, after picking the wrong card and triggering a bomb, the player can get out of the predicament by paying a certain amount of hard currency. This is the monetization angle of the feature, which is probably one of the reasons for its success.

The game designers at Rovio did a good balancing job so that a player always walks with enough items to justify participation in this minigame. There are several other mechanisms that ensure the fairness of the gameplay. Some of the floors are so-called Jackpot floors and do not have a risk of finding a bomb.

The Tower of Fortune is accessible for free once a day, but players can buy tickets for additional access as an IAP (see Fig. 10.15).

Paid Progress Plan

Another similar feature is known as a Paid Progression Plan. In a way, this feature works in the opposite way of the piggy bank. In the case of the piggy bank, the player gradually earns the currency that he can get in one big payout. In a paid progression plan, the player buys a certain amount of currency that will get delivered in installments.

The paid progression plans work in the following way. At some convenient point in the game, the player is served with an offer. The player can make a special in-app purchase

Fig. 10.15 Buying access to Tower of Fortune

involving a certain amount of virtual currency or other resources. A part of that amount is delivered to the player instantly after the purchase. The remainder is delivered gradually whenever the player reaches a certain milestone in the game (see Fig. 10.16).

For example, upon reaching level 10, the player can buy a paid progression plan that includes 10 000 gold coins. The player receives 1000 coins instantly. The player receives

Fig. 10.16 An example of a paid progression plan in family farm adventure by Century Games Pte. Ltd.

an additional 2,000 points upon reaching level 20, 3,000 upon reaching level 30, and 4,000 when reaching level 40.

The currency is again sold at a heavy discount. On the surface, this is a direct monetization feature, but its true purpose is to reinforce player retention. The player who has invested in a paid progression plan is more likely to remain with the game, at least until the last installment of the payout has been collected.

Glossary of Terms

Gameplay Terms

DPS

Damage Per Second—The amount of damage that some gameplay element, for example, a weapon, a character, or a spell can cause to opponents or targets during one second.

Mob

This is another name for computer-controlled non-playable characters of entities. It is more usually applied to characters that act as enemies or obstacles in games as opposed to characters with which players have verbal interaction. Good examples of mobs are Creepers or animals in Minecraft.

NPC

Non-Playable Characters—computer-controlled characters with which the player can come in contact inside the video game.

PvP

Player vs Player—a style of gameplay where the primary objective of each player involves overcoming other players, i.e. human participants.

PvE

Player vs Environment—a style of gameplay where the primary objective of a player is to overcome the obstacles posed by the game system, i.e. the environment.

TTK

Time to Kill—An amount of time it takes to kill another player in a game. In many games, characters have a property similar to health. If a character is hit by some kind of

© The Editor(s) (if applicable) and The Author(s), under exclusive license
to Springer Nature Switzerland AG 2024
S. Stanković, *Game Design for Free-to-Play Live Service*, Synthesis Lectures on Image,
Video, and Multimedia Processing, https://doi.org/10.1007/978-3-031-56156-6

weapon, a certain amount of health is deduced. Each such event takes some amount of time. Time to kill is the total time required to reduce the character's health to zero.

Game Genres

4X

Explore, Expand, Exploit, Exterminate—a sub-genre of strategy games best exemplified by the Civilization franchise. The emphasis in these games is on an exploration of the environment, expansion of an area under the player's control, exploitation of resources, and only eventually on the extermination of the opponents. On mobile platforms, this type of game is exemplified by games such as Game of War.

FPS

First-Person Shooter—a type of game in which the player controls a single character. Most often the gameplay revolves around some type of combat, usually with ranged weapons. The point of view of the game is from the perspective of the player's avatar, i.e. first-person perspective. The concept was pioneered in the early 90. By titles such as Wolfenstein 3D, Doom, and Quake. This is one of the most common types of games currently. Examples are numerous: Call of Duty, Battlefront, Far Cry, Apex Legends, Pub-G, Destiny, etc.

MMO

Massively Multiplayer Online game—a type of video game that can be played simultaneously by a multitude of people. These games are usually presented to users as persistent virtual worlds. This type of game is exemplified by games such as World of Warcraft, EVE Online, etc. This abbreviation can appear as a prefix in other abbreviations always denoting the same multiplayer aspects of the game.

MOBA

Multiplayer Online Battle Arena—a type of game exemplified by titles such as League of Legends or Vainglory. The gameplay is usually combat between two relatively small teams of players in a relatively confined area, i.e. arena. The objective of the game is typically the destruction of the opponent's main structure, a tower, or a fort for example. The teams consist of characters with various abilities. Characters are usually designed in a way so that the combination of their individual abilities offers synergy within the team. The terrain is usually constructed to offer a relatively easy passage through several corridors, known as lanes while the rest of the terrain is a mix of various types of obstacles.

RPG

Role Playing Games—a style of games where the accent is on character development. A player may control one or more of the virtual characters. Each of the characters is characterized by a series of properties, some of them numeric. These properties represent various qualities of the character, such as stamina, speed, health, etc. The properties have a direct effect on the gameplay. The player can gradually upgrade the properties of each of the characters under his control.

RTS

Real-Time Strategy game—a type of game in which the player is controlling an array of units. The player can issue commands to units in real-time. The gameplay involves combat between two human players of a human and a computer. These type of games was pioneered by titles such as Dune II. The most prominent examples are Warcraft, Command and Conquer, Total War, and Total Annihilation franchises.

TBS

Turn-Based Strategy—a type of game in which two or more players, either human or computer-controlled, fight each other in a game of strategy. Each side in the combat takes its own turn to move one or more of its units. Similar to the game of chess. The best examples of this genre of games are the Civilization franchise, Europa Universalis and Advanced Wars.

Game Design Terms

CCG

Card Collecting Game—a type of game, or a metagame structure in which the player is focused on collecting and upgrading a set of virtual or real cards. Good examples of these types of games are Magic the Gathering in the physical world or Hearthstone among video games. This type of metagame structure is used also in a wide variety of game genres, see for example Clash Royale.

Core Game, Core Gameplay, Core Interaction, or Core

This is a set of actions that the player will perform and a set of goals that the player will try to reach in every (or almost every) gameplay session. It is what is usually recognized as the basic gameplay. For example, jumping on platforms and over obstacles is the core gameplay of the Super Mario Bros game. Kicking the ball with your feet and trying to score a goal is the core gameplay of Soccer. All other aspects of the game can be considered a part of the Metagame.

FTUE

First-Time User Experience—also known as user onboarding, is the initial part of the game, everything that happens between the player and the game during the first play session or even several initial play sessions.

FPS

Frames Per Second—the number of frames that computer hardware can generate and display in one unit of time. This is one of the most common measures of gaming hardware performance.

HUD

Head-Up Display—in 3D games this is the part of the UI which consists of 2D elements superimposed over the 3D world. In a way similar to HUDs used in military jet fighters.

Magic Circle

A boundary that separates the game world from the real world. The game world is the place where the gameplay takes place Within the magic circle normal laws, patterns of behavior, and social conventions are temporarily suspended and replaced with the rules of the game.

Metagame, or Meta

These are all aspects of the game that are not part of the core gameplay. For example, a single level of the Super Mario Bros game is its core game. Everything else, including the overworld and the leaderboard, is a part of the metagame. Metagame can include level progression, various gameplay modes, character building, unlocking and upgrading logic, lore, a system of leagues or tournaments, etc. In a broader sense metagame can include activities that players undertake outside of the game itself. For example, organizing clan discussions via 3rd party software, or trading on 3d party websites can sometimes be considered a metagame.

Minute to Minute Interaction

This is another term used to describe the Core Gameplay.

MTX

Microtransaction Store—in free-to-play games this is the part of the game UI through which a player might perform purchases of virtual items, i.e. exchange of real-life money to some type of virtual items.

RPS

Rock, Paper, Scissors—a game design pattern where gameplay elements, items, characters, etc. are divided into three or more categories. Each category is superior in some aspects to one of the remaining two categories and inferior to the other one. The name of the pattern is derived from the classic playground game.

UX

User Experience—the term that denotes the way a user perceives the product. In the case of games, this is the holistic view of how a player perceives the game. In a more narrow sense, it is the player's perception of the UI of the game, its ease of use, functionality, etc.

UI

User Interface—the part of the software that facilitates the interaction between the player and the game. In a more narrow sense, it is a set of audiovisual elements, icons, menus, and static screens that players use to navigate through the game's software. The UI also includes audiovisual elements that convey some information about the state of the game to the player, i.e. status bars, score counters, life counters, etc.

XP

Experience Points—in some games this is a quantitative measure of a player's experience, i.e. the amount of time or effort that the player has spent in the game. Typically, a player will be granted some amount of experience points for performing certain actions within the game. For example, killing an opponent, finding a new item of a new type, uncovering a secret location, or simply chopping a virtual tree. XP can be used as an unlocking criterion for other game functions or features. For example, a player can be required to reach a certain level of XP in order to unlock a specific powerful weapon or to gain access to a location.

Business Terms

AAA

Usually pronounced as Triple-A is a business term denoting games that have made more than in cumulative revenue that is above a certain threshold. Over history, this threshold has been steadily moving upwards. In a more general sense, it refers to the top-of-the-line games. These games have very high production values, accompanied by very high development and marketing budgets. Consequently, they have the corresponding revenue expectations.

ARPU

Average Revenue Per User—total revenue of a game collected in a given time period divided by the number of users. This is one of the most common KPIs used to evaluate the performance of games run as an online service, very common in F2P games.

ARPPU

Average Revenue Per Paying User—similar to ARPU only calculated taking into account just the paying users in contrast to all active users.

ARPS

Average Revenue Per Spender—see ARPPU.

CARPU

Cumulative Average Revenue Per User—how much revenue in total and on average has an individual player contributed over some time period (e.g. D7 CARPU is CARPU over 7 days). It is calculated for example for all players who installed the game on the same day.

Churn

An act of a player abandoning a game.

Churn Point

A place within the game at which the player abandons a game. Often it can be an indication of the cause of a player's churn.

Churn Rate

The percentage of players that have abandoned a game up until a particular point. The inverse of retention rate.

Conversion Rate

A percentage of the total number of unique users that make at least one IAP purchase within the game.

CPI

Cost Per Install—This is one of the metrics used to evaluate the performance of user acquisition. It represents the cost of the acquisition of a single new user of an application or a game. It is calculated by dividing the total cost of the marketing campaign by the total number of users that have installed the game after seeing the ad.

D1/D2/D7/D30

Day 1/2/7/30—abbreviation denoting the particular number of days since some event. In the context of games that are run as a service, especially free-to-play games, this denotes a specific number of days after a user has installed the game. The Retention Rate of the game is calculated usually using these days as milestones. For example, the D7 retention rate is the number of players that still play the game one week after installing it divided by the total number of players that installed the game.

DAU

Daily Active Users—number of people, unique users that interact with a game during one day. This is one of the most common quantitative measures of the performance of a game, especially games that are run as online services. DAU is often used as a KPI.

EOD

End of Day—a deadline for something, implying that the deliverable should be ready by the end of standard working hours.

EOY

End of Year—a deadline, implying that the deliverable should be ready by the end of the year. The year can be a calendar year ending on 31 December or a fiscal year. Fiscal years can end on different dates depending on the context. For example, the fiscal year of the US government is the period beginning 1 October and ending 30 September the following year.

F2P

Free-to-play—a business model employed by some games. The game is distributed for free. Usually as a digital download. Players can play the game without the need to pay for the privilege. The game is usually financed via optional purchases or ad placements. This business model is especially popular with games on mobile platforms.

HD

High Definition—The term usually applied to the quality of graphics of a video game. It refers to either screen resolution or a certain FPS value. In a broader sense, it is used for all games that run on platforms that support such visuals, namely consoles and PC, as opposed to mobile games. In this context, the actual video performance characteristics are not important.

IAP

In-Application Purchase—a monetary transaction that takes place within an application. This is the way most microtransactions within F2P games on mobile platforms are implemented.

KPI

Key Performance Indicator—a quantitative measure that is used to evaluate the success of a particular game. Games that operate as online services can use a range of such measures including DAU, MAU, ARPU, ARPS, etc.

LTV

Lifetime Value—a sum total of all the money that an individual player has spent in the game from installation until churn.

MAU

Monthly Active Users— number of people, unique users that interact with a game during one month. This is one of the most common quantitative measures of the performance of a game, especially games that are run as online services. DAU is often used as a KPI.

Retention Curve

The retention rate for various days is plotted on a graph.

Retention Rate

One of the common quantitative measures used to estimate the performance of a game run as a service, especially free-to-play games. It is calculated as the number of users still interacting with a game after a certain number of days divided by the total number of people that had installed the game. For example, the D7 retention rate is the number of players that still play the game one week after installing it divided by the total number of players that installed the game. This is one of the most common KPIs.

Q1/Q2/Q3/Q4

Quarter 1/2/3/4—each year (both calendar and fiscal) is divided into four quarters each consisting of three months.

YoY

Year Over Year—a way of comparing two quantitative measures on an annual basis. It is commonly used in finance and business. For example, the revenue in the 3rd quarter in the current year can be compared to the revenue in the 3rd quarter of the previous year.

UA

User Acquisition—is an act of acquisition of new users for a game, an application, a service or a platform. In a more narrow sense, it includes all active efforts organized with this goal in mind, for example, marketing campaigns, promotions, offers, etc.

PBM

Performance-Based Marketing—a marketing model in which the performance of a particular advertising campaign can be monitored directly. This usually refers to digital advertising platforms, where technology permits the gathering of data about user behavior indicative of marketing performance. This can include the number of clicks on the ad, the percentage of players that have watched video ads until the end, etc. This data can further be used to modify the marketing strategy, justify the usage of funds, evaluate the performance of particular visuals, etc.

Team Roles

CM

Community Manager—a person in charge of communication and engagement with the player community of a particular game or gaming studio. A sort of PR of the game or the studio. This person is usually in charge of the social media accounts, such as Discord Servers, Forums, Web Site, Subreddits, Instagram, Twitter, Facebook Pages, Twitch, TikTok, etc., etc.

CS

Customer Support—these are the people corresponding directly with the customers. They are the ones usually handling the direct complaints and requests submitted by individual players, via specific official contact channels, such as contact emails, official forums, specific in-game features designed for this purpose etc.

EP

Executive Produces—Key vision holder of the game. This role is supposed to formulate the vision of the game and the strategic direction of the live service; to communicate this vision to all the team members and to guide, oversee and facilitate the operations of the team.

GM

Game Manager—usually the most senior person in charge of a project. Quite often this person is the main vision holder of the game. The requirements of the role may vary from team to team, but in general this person is supposed to determine, organize and oversee the team, facilitate the development of the game and the operations of the live service.

GGM

Game General Manager—pretty much the same as the above.

PM

Product Manager—the person usually in charge of the business side of the game development and live service operations.

QA

Quality Assurance—these are the people responsible for testing the implementation of the game. They are supposed to play the game with specific attention to newly added features and content, spot bugs, and report bugs. After the bugs have been fixed by the developers, QA is supposed to test the same bits of the game all over to verify that the bugs have been fixed correctly.

Bibliography

Aaronson, S. (2014). Quantum Randomness. American Scientist, 102(4), 266. https://doi.org/10.1511/2014.109.266

ActivePlayer.io. (2023). Top 15 Most Popular PC Games of 2023 - The Game Statistics Authority. ActivePlayer.io. Retrieved July 7, 2023, from https://activeplayer.io/top-15-most-popular-pc-games-of-2022/

Adams, E. (2014). Fundamentals of Game Design. New Riders.

Alexander, L., & Matthews, M. (2009). Analysis: GameStop Profit Margin On Used Almost 50%. Game Developer. Retrieved July 6, 2023, from https://www.gamedeveloper.com/pc/analysis-gamestop-profit-margin-on-used-almost-50-

Ambrose, S. H. (1998, June). Late Pleistocene human population bottlenecks, volcanic winter, and differentiation of modern humans. Journal of Human Evolution, 34(6), 623–651. ScienceDirect. https://doi.org/10.1006/jhev.1998.0219

Ambrose, S. H. (1998, June). Late Pleistocene human population bottlenecks, volcanic winter, and differentiation of modern humans. Journal of Human Evolution, 36(6), Pages 623–651. https://doi.org/10.1006/jhev.1998.0219

Ansoff, I. H. (1980, June). Strategic issue management. Strategic Management Journal, 1(2), 131–148. Wiley Online Library. https://doi.org/10.1002/smj.4250010204

Aponte, M.-V., Levieux, G., & Natkin, S. (2009). Scaling the Level of Difficulty in Single Player Video Games. International Conference on Entertainment Computing, 24–35. 0.1007/978-3-642-04052-8_3

Attwood, A. S., Scott-Samuel, N. E., Stothart, G., & Munafò, M. R. (2012, August 17). Glass Shape Influences Consumption Rate for Alcoholic Beverages. Plos One. https://doi.org/10.1371/journal.pone.0043007

Baker, R. (2021, January 25). Player Personas in Mobile Games: Spending and Personality. Appreciation Engine. Retrieved July 10, 2023, from https://get.theappreciationengine.com/2021/01/25/mobile-game-player-personas/

Bartle, R. A. (2004). Designing virtual worlds. New Riders.

Blacker, A. (2023, January 4). Worldwide and US Download Leaders 2022. Apptopia Blog. Retrieved July 7, 2023, from https://blog.apptopia.com/worldwide-and-us-download-leaders-2022#Games

Brassey, J., Kuo, G., Murphy, L., & van Dam, N. (2019, February 15). Shaping individual development along the S-curve. Insights on People and Organizational Performance. https://www.mckinsey.com/capabilities/people-and-organizational-performance/our-insights/shaping-individual-development-along-the-s-curve/

S. Stanković, *Game Design for Free-to-Play Live Service*, Synthesis Lectures on Image, Video, and Multimedia Processing, https://doi.org/10.1007/978-3-031-56156-6

Buchanan, J. M. (1991). Opportunity Cost. In P. Newman, J. Eatwell, & M. Milgate (Eds.), The World of Economics (pp. 520–525). Palgrave Macmillan UK. https://doi.org/10.1007/978-1-349-21315-3_69

Butler, S., & Gloor, J. (2023, March 28). How to Create ChatGPT Personas for Every Occasion. How-To Geek. Retrieved July 12, 2023, from https://www.howtogeek.com/881659/how-to-create-chatgpt-personas-for-every-occasion/

Bycer, J. (2018). 20 Essential Games to Study. CRC Press.

Byrne, R. M. J. (2005). The rational imagination: how people create alternatives to reality. MIT Press.

Caillois, R. (1957). Les jeux et les hommes: le masque et le vertige. Gallimard.

Campbell, J. (2008). The hero with a thousand faces. New World Library.

Carcasole, D. (2021, April 21). Netflix to spend $17B on production in 2021. CGMagazine. Retrieved July 5, 2023, from https://www.cgmagonline.com/news/netflix-spend-17-billion-production/

Cheesemeister. (2022, July 14). "Mario games teach us that even if something is essentially the same, psychologically it can be completely different. This example is very easy to understand." Twitter. https://twitter.com/Cheesemeister3k/status/1547440825420099586?s=20

Clarysse, P. (2014, September 8). Rovio: Leading the way on gender balance. GamesIndustry.biz. Retrieved July 11, 2023, from https://www.gamesindustry.biz/rovio-leading-the-way-on-gender-balance-1

Clement, J. (2022, December 1). U.S. PUBG Mobile users by gender 2019. Statista. Retrieved July 11, 2023, from https://www.statista.com/statistics/988128/pubg-mobile-players-us-gender/

Clement, J. (2023, June 28). Video Game Industry - Statistics & Facts. Statista. Retrieved July 6, 2023, from https://www.statista.com/topics/868/video-games/#topicOverview

Constable, G. (2009, September 29). Why do people buy virtual goods? (on motivations and compulsions) | giffconstable.com. giffconstable.com. Retrieved July 3, 2023, from https://giffconstable.com/2009/09/why-do-people-buy-virtual-goods-on-motivations-and-compulsions/

Costa, P. T., Terracciano, A., & McCrae, R. R. (2001). Gender differences in personality traits across cultures: robust and surprising findings. Journal of Personality and Social Psychology, 81(2). https://doi.org/10.1037//0022-3514.81.2.322

Crawford, C. (1984). The Art of Computer Game Design. Osborne/McGraw-Hill.

Csikszentmihalyi, M. (1990). Flow: The Psychology of Optimal Experience. Harper & Row.

Csikszentmihalyi, M. (2008). Flow: The Psychology of Optimal Experience. HarperCollins.

Curry, D. (2023, January 9). Home App Data Among Us Revenue and Usage Statistics (2023). Business of Apps. Retrieved July 4, 2023, from https://www.businessofapps.com/data/among-us-statistics/

Desarbo, W. S., Hausman, R. E., & Kukitz, J. M. (2007, november). Restricted principal components analysis for marketing research. Journal of Modelling in Management, 2(3), 305–328. https://doi.org/10.1108/17465660710834471

Dewanto, G. W., & Tiatri, S. (2021). Attitude Towards Video Game Genre and Five-Trait Personality. Advances in Social Science, Education and Humanities Research, 570.

Digman, J. M. (1990, February). Personality Structure: Emergence of the Five-Factor Model. Annual Review of Psychology, 41, 417-440. https://doi.org/10.1146/annurev.ps.41.020190.002221

Doran, G. T. (1981). There's a S.M.A.R.T. way to write management's goals and objectives. Management Review, 70, 35-36.

Douglas, K. M., Sutton, R. M., & Cichocka, A. (2019). Belief in Conspiracy Theories: Looking beyond gullibility. In J. P. Forgas & R. F. Baumeister (Eds.), The Social Psychology of Gullibility: Fake News, Conspiracy Theories, and Irrational Beliefs (pp. 61–76). Routledge. https://doi.org/10.4324/9780429203787-4

Draven, D. (2020, May 1). The 10 Most Controversial Vaporware Game Titles In History, Ranked. TheGamer. Retrieved July 23, 2023, from https://www.thegamer.com/vaporware-video-games-most-controversial

Duda, R. O., Hart, P. E., & Stork, D. G. (2000). Pattern classification (2nd ed.). Wiley.

Edwards, B. (2015, January 22). The Untold Story Of The Invention Of The Game Cartridge. Fast Company. Retrieved July 6, 2023, from https://www.fastcompany.com/3040889/the-untold-story-of-the-invention-of-the-game-cartridge

Edwards, B. (2021, August 24). What Is Shareware, and Why Was It So Popular in the 1990s? How-To Geek. Retrieved July 6, 2023, from https://www.howtogeek.com/728527/what-is-shareware-and-why-was-it-so-popular-in-the-1990s/

Eelectronic Software Association. (2022). Essential Facts About the Video Game Industry.

Elo, A. E. (1967). The Proposed USCF Rating System, Its Development, Theory, and Applications. Chess Life, 12(8), 242–247.

Emelyantseva, D. (2019, April 18). 5 Basic Steps in Creating Balanced In-Game Economy. Game Developer. Retrieved July 4, 2023, from https://www.gamedeveloper.com/design/5-basic-steps-in-creating-balanced-in-game-economy

Ende, M. v. d., Rohlfs, S., Yagafarova, A., Bas, P. d., Poort, J., Haffner, R., & Till, H. v. (2014). Estimating Displacement Rates of Copyrighted Content in the EU: Final Report. Publications Office of the European Union.

England, L. (2014, April 24). "The Door Problem" of Game Design. Game Developer. Retrieved July 6, 2023, from https://www.gamedeveloper.com/design/-quot-the-door-problem-quot-of-game-design#close-modal

England, L. (2014). Types of Designers. Game Developer. Retrieved July 9, 2023, from https://www.gamedeveloper.com/design/types-of-designers

English, T. (2020, March 8). You Could Download Video Games From the Radio in the 1980s. Interesting Engineering. Retrieved July 6, 2023, from https://interestingengineering.com/science/you-could-download-video-games-from-the-radio-in-the-1980s

Fairfax, Z. (2020, March 11). Fortnite: How to Gift A Battle Pass. Screen Rant. Retrieved July 3, 2023, from https://screenrant.com/fortnite-gift-battle-pass/

Falk, A., & Hermle, J. (2018, October 19). Relationship of gender differences in preferences to economic development and gender equality. Science, 362(6412). https://doi.org/10.1126/science.aas9899

Fehr, E., & Gächter, S. (2000). Fairness and Retaliation: The Economics of Reciprocity. Journal of Economic Perspectives, 14(3), 159-181. https://doi.org/10.1257/jep.14.3.159

Fullerton, T. (2018). Game Design Workshop: A Playcentric Approach to Creating Innovative Games. CRC Press.

Garcia-Valera, F., Kazimierczak, M., Arias Burgos, C., & Vajsman, N. (2021). Online Copyright Infringement in the European Union: Music, Films and TV (2017-2020), Trends and Drivers. Publications Office of the European Union.

Glen, S. (n.d.). IID Statistics: Independent and Identically Distributed Definition and Examples. Statistics How To. Retrieved July 3, 2023, from https://www.statisticshowto.com/iid-statistics/

Goff, M. (2022, June 2). Software Piracy and License Misuse Stat Watch. Revenera. Retrieved July 7, 2023, from https://www.revenera.com/blog/software-monetization/software-piracy-stat-watch/

Goldberg, L. R. (1993). The structure of phenotypic personality traits. American Psychologist, 48(1), 26–34. https://doi.org/10.1037/0003-066X.48.1.26

Goodwin, K. (2009). Designing for the Digital Age: How to Create Human-Centered Products and Services. Wiley.

Haight. (1967). Handbook of the Poison Distribution. John Wiley & Sons Canada, Limited.

Hamari, J., & Keronen, L. (2017, July). Why do people buy virtual goods: A meta-analysis. Computers in Human Behavior, 71(1), 59–69. j.chb.2017.01.042.

Haselton, M. G., & Nettle, D. (2015). The Evolution of Cognitive Bias. In D. M. Buss (Ed.), The Handbook of Evolutionary Psychology, 2 Volume Set (pp. 724–746). Wiley. https://doi.org/10.1002/9781119125563.evpsych241

Herbrich, R., Minka, T., & Graepel, T. (2007, January). TrueSkill(TM): A Bayesian Skill Rating System. Advances in Neural Information Processing Systems, 20. https://www.microsoft.com/en-us/research/publication/trueskilltm-a-bayesian-skill-rating-system/

Hirsh-Pasek, K., Golinkoff, R. M., & Eyer, D. (2004). Einstein Never Used Flash Cards: How Our Children Really Learn--and Why They Need to Play More and Memorize Less. Harmony/Rodale.

Hrodey, M. (2020, May 30). Meet the man who invented microtransactions years before Oblivion's horse armour. PCGamesN. Retrieved July 7, 2023, from https://www.pcgamesn.com/first-game-with-microtransactions

https://app.sensortower.com/usage-intel/demographics?selected_tab=demographics&os=ios&country=all&period=day&start_date=2015-10-01&end_date=2023-07-06&saa=com.rovio.angrybirdsstella&sia=875251011

https://gameanalytics.com/blog/how-to-create-user-personas-as-unique-as-your-audience/

https://www.gamedeveloper.com/blogs/video-games-and-personality-traits-a-deep-dive-into-the-science-of-gaming-preferences

https://www.theesa.com/resource/2022-essential-facts-about-the-video-game-industry/

https://www.wiley.com/en-it/Pattern+Classification,+2nd+Edition-p-9780471056690

Huizinga, J. (1955). Homo Ludens: A Study of the Play-Element in Culture. Beacon Press.

Hussain, U., Jabarkhail, S., Cunningham, G. B., & Madsen, J. A. (2021). The dual nature of escapism in video gaming: A meta-analytic approach. Computers in Human Behavior Reports, 3.https://doi.org/10.1016/j.chbr.2021.100081

Jolliffe, I. T., & Cadima, J. (2016, April 13). Principal component analysis: a review and recent developments. philosophical Transactions of the Royal Society A: Mathematical, Physical and Engineering Sciences, 374(2065). https://doi.org/10.1098/rsta.2015.0202

Kachergis, G. (2023, May 3). Video Games and Personality Traits: A Deep Dive into the Science of Gaming Preferences. Game Developer. Retrieved July 11, 2023, from

Kahn, J. H., & Wright, S. E. (1980). Human Growth and the Development of Personality. Elsevier Science & Technology.

Katkoff, M. (2020, October 27). How Playrix' Township Became a Billion Dollar Game—Deconstructor of Fun. Deconstructor of Fun. Retrieved July 4, 2023, from https://www.deconstructoroffun.com/blog/2020/10/13/how-playrix-township-became-a-billion-dollar-game

Kent, S. L. (2001). The Ultimate History of Video Games. Three Rivers Press.

Knezovic, A. (2023). Design Home Analysis: How It Targets Millennial Women? - Udonis. Mobile Marketing. Retrieved July 11, 2023, from https://www.blog.udonis.co/mobile-marketing/mobile-games/design-home-analysis

Kong, L., & O'Connor, J. (2009). Creative Economies, Creative Cities: Asian-European Perspectives (L. Kong & J. O'Connor, Eds.). Springer Netherlands.

Koster, R. (2013). A Theory of Fun for Game Design. O'Reilly.

Landsberg, M. (2015). The Tao of Coaching: Boost Your Effectiveness at Work by Inspiring and Developing Those Around You. Profile Books.

Librande, S. (2010). One-Page Design. Stonetronix. Retrieved July 25, 2023, from http://stonetronix.com/gdc-2010/OnePageDesigns.ppt

Lidwell, W., Holden, K., & Butler, J. (2010). Universal Principles of Design, Revised and Updated: 125 Ways to Enhance Usability, Influence Perception, Increase Appeal, Make Better Design Decisions,. Rockport Publishers.

Liew, J. (2009, February 9). Why do People Buy Virtual Goods? - WSJ. The Wall Street Journal. Retrieved July 3, 2023, from https://www.wsj.com/articles/SB123395867963658435

Loerich, A. G. (2016, January 1). Learning Curves. Encyclopedia of Operations Research and Management Science, 871–874. https://doi.org/10.1007/978-1-4419-1153-7_526

Long, N. (2022, October 31). Zynga Helsinki closing, underperforming FarmVille 3 moves to Bengaluru. Mobilegamer.biz. Retrieved July 4, 2023, from https://mobilegamer.biz/zynga-helsinki-closing-underperforming-farmville-3-moves-to-bengaluru/

Lovell, N. (2013). The Curve: From Freeloaders Into Superfans : the Future of Business. Portfolio Pearson.

Luton, W. (2022, January 13). The Mathematics of Game Balance | Blog. UserWise. Retrieved July 6, 2023, from https://www.userwise.io/blog/the-mathematics-of-game-balance

MacQueen, J. (1967). Some methods for classification and analysis of multivariate observations. Proc. of Berkeley Symp. on Math. Statist. and Prob., 5(1), 281–297.

Maslow, A. H. (1943). A theory of human motivation. Psychological Review, 50(4), 370–396. https://doi.org/10.1037/h0054346

Mattioli, M. (2020). History of Video Game Distribution. IEEE Consumer Electronics Magazine, 1(1), 99. https://www.researchgate.net/publication/346358929_History_of_Video_Game_Distribution

Mattson, M. P. (2014). Superior pattern processing is the essence of the evolved human brain. Frontiers in Neuroscience, 8(1). https://doi.org/10.3389/fnins.2014.00265

Millidge, B., Seth, A., & Buckley, C. L. (n.d.). Predictive Coding: a Theoretical and Experimental Review. arXiv:CS. arXiv. https://doi.org/10.48550/arXiv.2107.12979 Focus to learn more

Minka, T., Cleven, R., & Zaykov, Y. (2018, March 22). TrueSkill 2: An improved Bayesian skill rating system. Microsoft. Retrieved July 5, 2023, from https://www.microsoft.com/en-us/research/publication/trueskill-2-improved-bayesian-skill-rating-system/

Mischel, W., & Ebbesen, E. B. (1970). Attention in delay of gratification. Journal of Personality and Social Psychology, 16(2), 329–337. https://doi.org/10.1037/h0029815

Mitchell, R. (2023, January 5). Gower Street Estimates 2022 Global Box Office Hit $25.9 Billion. Gower Street Analytics. Retrieved July 6, 2023, from https://gower.st/articles/gower-street-estimates-2022-global-box-office-hit-25-9-billion/

Morgenstern, O., & von Neumann, J. (1944). Theory of Games and Economic Behavior: 60th Anniversary Commemorative Edition. Princeton University Press.

Mulligan, M. (2023, March 16). Cover image for Recorded music market 2022 | Reality bites. MIDiA Research. Retrieved July 6, 2023, from https://midiaresearch.com/blog/recorded-music-market-2022-reality-bites

Neele, J., & Speetjens, R. (2022, January 24). Consumer trends in 2022: the subscription economy and the metaverse. Robeco. Retrieved July 5, 2023, from https://www.robeco.com/en-int/insights/2022/01/consumer-trends-in-2022-the-subscription-economy-and-the-metaverse

NewZoo. (2023, May). Most played PC games - Global | By MAU. Newzoo. Retrieved July 26, 2023, from https://newzoo.com/resources/rankings/top-20-pc-games

Nissen, T. (2004). From 102 to 915 million combinations. LEGO Corporate Communications.

Nixon, J. (2017). Football World: History of Football. Hachette Children's Group.

Oppenheimer, D. M., & Monin, B. (2009, August). The retrospective gambler's fallacy: Unlikely events, constructing the past, and multiple universes. Judgment and Decision Making, 4(5), 326–334. https://doi.org/10.1017/S1930297500001170

Pace, L. (2022, December 6). Floppy Disk Explained: Everything You Need to Know. History-Computer. Retrieved July 6, 2023, from https://history-computer.com/floppy-disk/

Pagel, M., Söbke, H., & Bröker, T. (2021, October). Using Multiplayer Online Games for Teaching Soft Skills in Higher Education. Serious Games. JCSG 2021. Lecture Notes in Computer Science, 12945. https://doi.org/10.1007/978-3-030-88272-3_20

Pelleg, D., & Moore, A. W. (2000). X-means: Extending K-means with Efficient Estimation of the Number of Clusters. ICML '00: Proceedings of the Seventeenth International Conference on Machine Learning, 727–734. https://doi.org/10.5555/645529.657808

Perreault, G., & Vos, T. (2019, June). Metajournalistic discourse on the rise of gaming journalism. New Media & Society, 22(1). https://doi.org/10.1177/1461444819858695

Petru, S. (2017, December 27). I remember. Differences between the Neanderthal and modern human mind. Documenta Praehistorica, 44(1). https://doi.org/10.4312/dp.44.25

Phaal, R., Farrukh, C. J.P., & Robert, D. R. (2004, February). Technology roadmapping—A planning framework for evolution and revolution. Technological Forecasting and Social Change, 71(1–2), 5–26. ScienceDirect. https://doi.org/10.1016/S0040-1625(03)00072-6

Philippou, C. (2021, October 16). Commentary: Why are the mega-rich buying up English Premier League football clubs? CNA. Retrieved July 7, 2023, from https://www.channelnewsasia.com/commentary/english-premier-league-newcastle-saudi-manchester-chelsea-money-2246836

Preply. (n.d.). Study Reveals the Most Annoying Corporate Jargon. Preply. Retrieved July 4, 2023, from https://preply.com/en/learn/best-and-worst-corporate-jargon

Quintais, J., & Poort, J. (2018). Global Online Piracy Study. Kluwer Copyright Blog.

Rapoport, A., & Chammah, A. M. (1965). Prisoner's Dilemma: A Study in Conflict and Cooperation. University of Michigan Press.

Rappaz, J., Catasta, M., West, R., & Aberer, C. (2018). Latent Structure in Collaboration: The Case of Reddit r/place. Proceedings of the International AAAI Conference on Web and Social Media, 12(1). https://doi.org/10.1609/icwsm.v12i1.15013

Rigby, S., & Ryan, R. (2004). The Player Experience of Need Satisfaction (PENS) [White paper]. immersyve.com. https://natronbaxter.com/wp-content/uploads/2010/05/PENS_Sept07.pdf

Rigby, S., & Ryan, R. M. (2011). Glued to Games: How Video Games Draw Us In and Hold Us Spellbound. Bloomsbury Academic.

Roberts, B. W., Walton, K. E., & Viechtbauer, W. (2006). Patterns of mean-level change in personality traits across the life course: A meta-analysis of longitudinal studies. Psychological Bulletin,, 132(1), 1-25. https://doi.org/10.1037/0033-2909.132.1.1

Rorty, A., & McLaughlin, B. P. (Eds.). (1988). Perspectives on Self-Deception. University of California Press.

Rothmann, S., & Coetzer, E. P. (2003, October 24). The big five personality dimensions and job performance. SA Journal of Industrial Psychology, 29(1). https://doi.org/10.4102/sajip.v29i1.88

Ruby, D. (2023, February 7). Fortnite Statistics For 2023 (Users, Revenue & Devices). Demand Sage. Retrieved July 11, 2023, from https://www.demandsage.com/fortnite-statistics/

Ryan, R. M., & Deci, E. L. (2000). Self-determination theory and the facilitation of intrinsic motivation, social development, and well-being. American Psychologist, 55(1), 68–78. https://doi.org/10.1037/0003-066X.55.1.68

Ryan, R. M., & Deci, E. L. (2018). Self-Determination Theory: Basic Psychological Needs in Motivation, Development, and Wellness. Guilford Publications.

Ryan, R. M., & Deci, E. L. (2020). Self-determination theory and the facilitation of intrinsic motivation, social development, and well-being. American Psychologist, 55(1), 68–78. APA PsyNet. https://doi.org/10.1037/0003-066X.55.1.68

Salminen, J., Vahlo, J., Koponen, A., Jung, S.-G., Chowdhury, S. A., & Jansen, B. J. (2020). Designing Prototype Player Personas from a Game Preference Survey. Conference on Human Factors in Computing Systems. https://doi.org/10.1145/3334480.3382785

Sawicki, M., & Moody, J. (2020). A brief history of computer graphics. In Filming the Fantastic with Virtual Technology: Filmmaking on the Digital Backlot (p. 36). Taylor & Francis Group. https://www.taylorfrancis.com/chapters/edit/https://doi.org/10.4324/9780429331282-3/brief-history-computer-graphics-mark-sawicki-juniko-moody

Schell, J. (2008). The art of game design. Taylor & Francis.

Schell, J. (2019). The Art of Game Design: A Book of Lenses. CRC Press.

Schittkowski, K. (2002). Numerical data fitting in dynamical systems. Kluwer Academic Publishers.

Scully, E. (2020, August 1). Best Games Made By One Person - The Top Video Games Created By a Single Developer. Career Karma. Retrieved July 12, 2023, from https://careerkarma.com/blog/games-made-by-one-person/

SensorTower. (n.d.). Angry Birds Stella Usage Intelligence - Demographics. SensorTower. Retrieved July 11, 2023, from

Sherif, M., Taub, D., & Hovland, C. I. (1958). Assimilation and contrast effects of anchoring stimuli on judgments. Journal of Experimental Psychology, 55(2), 150–155. https://doi.org/10.1037/h0048784

Short, T. X., Hurd, D., Forbes, J., Diaz, J., Ordon, A., Howe, C., Eiserloh, S., & Cook, D. (2017, January 1). Group Report: Coziness in Games: An Exploration of Safety, Softness, and Satisfied Needs. Project Horseshoe. Retrieved July 3, 2023, from https://www.projecthorseshoe.com/reports/featured/ph17r3.htm

Social Security Administration of US. (n.d.). Popular Baby Names by Decade. SSA. Retrieved July 10, 2023, from https://www.ssa.gov/oact/babynames/decades/index.html

Solsten. (n.d.). Traits - AI-Powered Player Intelligence Platform. Solsten. Retrieved July 12, 2023, from https://www.solsten.io/traits/

Statista. (2023, May 2). U.S. mean disposable household income by age 2021. Statista. Retrieved July 11, 2023, from https://www.statista.com/statistics/980324/us-mean-disposable-household-income-age/

Stefańska, M., & Śmigielska, G. (2021, September 21). Impulse Purchase in Virtual Environment and Price Sensitivity of Young Consumers: Results of Empirical Research. Economy and Market Communication Review, 19(1). https://doi.org/10.7251/EMC2001008S

Stoll, J. (2023, April 20). Netflix: quarterly revenue 2023. Statista. Retrieved July 5, 2023, from https://www.statista.com/statistics/273883/netflixs-quarterly-revenue/

Swedell, L. (2012). Primate Sociality and Social Systems. Nature Education Knowledge, 3(10), 84.

Szyller, S. (2022, March 15). sebszyller.com. sebszyller.com. Retrieved July 10, 2023, from https://sebszyller.com/blog/2022/datamarketplaces

Unsplash. (n.d.). 500+ Persona Pictures | Download Free Images on Unsplash. Unsplash. Retrieved July 12, 2023, from https://unsplash.com/s/photos/persona

Vahlo, J., & Koponen, A. (2018). Player Personas and Game Choice. In N. Lee (Ed.), Encyclopedia of Computer Graphics and Games. Springer International Publishing. https://doi.org/10.1007/978-3-319-08234-9_149-1

Vlerick, M. (2020). Explaining human altruism. Synthese volume, 199, 2395–2413. Springer Link. https://doi.org/10.1007/s11229-020-02890-y

Ward, M. (2016, December 9). What do Prince Charles and Ozzy Osbourne have in common? BBC. Retrieved July 11, 2023, from https://www.bbc.com/news/technology-37307829.amp

Wardle, S. G., Taubert, J., & Baker, C. I. (2020, August 9). Rapid and dynamic processing of face pareidolia in the human brain. Nature Communications, 11, 4518. https://doi.org/10.1038/s41467-020-18325-8

Wenk, G. L. (2023, January 7). How Social Isolation Affects the Brain. Psychology Today. Retrieved July 26, 2023, from https://www.psychologytoday.com/intl/blog/your-brain-on-food/202301/how-social-isolation-affects-the-brain

Whitmore, J. (2017). Coaching for Performance Fifth Edition: The Principles and Practice of Coaching and Leadership UPDATED 25TH ANNIVERSARY EDITION. Mobius.

Wijman, T. (2022, July 26). Global Games Market Report. Newzoo. https://newzoo.com/resources/trend-reports/newzoo-global-games-market-report-2022-free-version

Wijman, T. (2023). Global Games Market Report 2023. Newzoo. https://newzoo.com/resources/blog/newzoos-game-market-trends-to-watch-in-2023-part-1

Wittgenstein, L. (1953). Philosophical Investigations: The German Text, with a Revised English Translation 50th Anniversary Commemorative Edition (G. E. M. Anscombe, Trans.). Wiley.

Wolf, M. J. P. (Ed.). (2012). Before the Crash: Early Video Game History. Wayne State University Press.

Xiangyi, F. (2016). Horror movie aesthetics: how color, time, space and sound elicit fear in an audience [Masters theses]. Northeastern University, Boston, Massachusetts. https://doi.org/10.17760/D20211378

Yancy, L. (2019, May 28). 'If You're Playing EVE Online You Basically Already Have An MBA,' Says Player Who Built His Own Company [UPDATE]. Kotaku. Retrieved July 26, 2023, from https://kotaku.com/if-you-re-playing-eve-online-you-basically-already-have-1835048929

Yu, Y. (2017, February 23). Steam – the gaming platform before there were platforms - Digital Innovation and Transformation. Digital, Data, and Design Institute at Harvard. Retrieved July 6, 2023, from https://d3.harvard.edu/platform-digit/submission/steam-the-gaming-platform-before-there-were-platforms/

Yunxiang, Y. (2020). Gifts. In The Open Encyclopedia of Anthropology (p. 1). Division of Social Anthropology University of Cambridge. https://doi.org/10.1006/jhev.1998.0219

Zahavi, A. (1975, September). Mate selection—A selection for a handicap. Journal of Theoretical Biology, 53(1), 205–214. Sciencedirect. 0022–5193(75)90111–3

Zenn, J. (2017, May 24). How to Create User Personas in your Analytics Reports. GameAnalytics. Retrieved July 10, 2023, from

Zimmerman, E., Salen, K., & Salen Tekinbas, K. (2003). Rules of Play. Books24x7.com.